Fundamentals of Meteorology

Prentice-Hall, Inc., Englewood Cliffs, New Jersey 07632

Fundamentals of

LOUIS J. BATTAN

Director
Institute of Atmospheric Physics
University of Arizona

Library of Congress Cataloging in Publication Data

BATTAN, LOUIS J
   Fundamentals of meteorology.

   Bibliography: p. 293
   Includes index.
   1. Meteorology. I. Title.
QC861.2.B38   551'.5   78-7799
ISBN 0-13-341131-1

FUNDAMENTALS OF METEOROLOGY
Louis J. Battan

Printed in the United States of America

10 9 8 7 6 5 4 3 2 1

Editorial/production supervision by *Karen J. Clemments*
Cover/interior design by *Mark A. Binn*
Page layout by *Georgina Ottaviano*
Manufacturing buyer: *Trudy Pisciotti*
Cover photograph supplied by the *National Oceanic and Atmospheric Administration*

PRENTICE-HALL INTERNATIONAL, INC., *London*
PRENTICE-HALL OF AUSTRALIA PTY. LIMITED, *Sydney*
PRENTICE-HALL OF CANADA, LTD., *Toronto*
PRENTICE-HALL OF INDIA PRIVATE LIMITED, *New Delhi*
PRENTICE-HALL OF JAPAN, INC., *Tokyo*
PRENTICE-HALL OF SOUTHEAST ASIA PTE. LTD., *Singapore*
WHITEHALL BOOKS LIMITED, *Wellington, New Zealand*

Contents

4 The Winds *63*

5 Atmospheric Stability and Vertical Motions *80*

6 Planetary Patterns of Air Motions *101*

7 Air Masses, Fronts, and Cyclones *121*

8 Clouds, Rain, and Snow *147*

9 Severe Storms *181*

10 Atmospheric Optics and Acoustics 206

11 Climates of the Earth 224

12 Applications 258

Preface

THE WEATHER AND CLIMATE AFFECT LIFE on the earth in many important ways—some direct and obvious, others indirect and subtle. Violent storms, dealing death and destruction, clearly fall into the first category. The results of droughts in distant places sometimes are not as readily discernable, even though their consequences to society can be profound. For example, an extensive drought in the Soviet Union in 1972 caused that country to buy large quantities of wheat from the United States. As a consequence the price of bread increased for all Americans and for the citizens of many other countries as well.

One of the purposes of this book is to help students perceive the role of the atmosphere in human affairs. The chief goal is to introduce college students to meteorology and climatology.

This volume is designed primarily for those with little training in mathematics, physics and chemistry. For this reason, there are few mathematical formulations in the main body of the text. A number of the more significant equations of meteorology are given in footnotes for those individuals who are interested. Since they are not essential in an introductory course, readers who are not fond of equations are invited to ignore these footnotes.

After a survey of meteorology a student should have a better understanding of the atmosphere—what we know and what still remains to be learned. It should be possible to look at the sky with a greater appreciation of the nature of clouds and storms and associated phenomena. A student should be in a better position to see how the weather and climate are related to his or her particular area of interest. We hope some individuals will be inspired to learn more about this challenging subject and perhaps make a career of the atmospheric sciences.

LOUIS J. BATTAN

Fundamentals of Meteorology

CHAPTER 1

Introduction

BEFORE LAUNCHING into a detailed discussion of the science and practice of meteorology, we believe that for students exposed to this subject for the first time, it would be helpful to give a brief overview of what is to come. In reading this introductory chapter, we advise that you skim through it rapidly. There is no need to underline or try to memorize anything. The purpose is to introduce you to the entire subject in a general way, to give you an understanding of the uniqueness of the earth's atmosphere, and to impress you with the challenges of weather and climate. It is hoped that this chapter will make you aware of how the atmosphere affects life on earth and will make you eager to learn more about it.

The planet earth is unique in many important ways. It has expansive oceans and a substantial atmosphere made up of those gases which have allowed life, as we know it, to evolve and continue to exist. Only the earth, among the heavenly bodies orbiting the sun, has these properties. In view of the countless millions of other suns in the universe, most or all of which may have their own planets, it has been speculated that there may be many other civilizations, some much more advanced than our own. Astronomers have begun to listen for radio signals from outer space proving the existence of humanly creatures in the great beyond among the distant stars. The detection of such signals would be an event of immense importance.

Science fiction writers long have written about space travel and colonization of the moon and the planets. Not too many years ago such stories seemed farfetched and surely not likely to be achieved in this century. But on July 20, 1969, Neil Armstrong set foot on the moon and space travel became a reality.

Space vehicles have observed the properties of the nearby planets—Venus, Mars, Jupiter in particular. In a few instances, instruments have descended through the faraway planetary atmospheres and telemetered back measurements of gaseous composition, temperature, pressure, and wind. More often, specialized telescopes and cameras have observed and photographed the surfaces and atmospheres as spacecrafts passed relatively close to the planets. We know that the ones in our solar system have environments hostile to the life thriving on the earth.

The atmospheres of Venus and Mars are mostly carbon dioxide while that of Jupiter is mostly hydrogen. Oxygen, that crucial gas on which life

2

depends, is almost a stranger in the atmospheres of these astronomical neighbors. The temperatures on these planets are hostile to living creatures. Venus is too hot while Mars and Jupiter are too cold.

Someday people will travel to nearby planets as they have to the moon, but it will be necessary to live within complicated life-supporting cocoons. Proposals dealing with space cities circling the earth and with immigration to other planets are fascinating, but it is certain that for a long time into the future, the earth will be home for the human race.

When the world's population was small, the supplies of clean air, water, and food and the mineral resources in the ground seemed limitless. There appeared to be little need to be thrifty in the use of these necessities; inefficiencies became a way of life, particularly in the more advanced countries of the world. In the last few decades it has become increasingly clear that the earth's resources are not inexhaustible.

An examination of factors governing the supplies of food and water and the use of fuels brings into immediate focus the vital parts played by weather and climate. Prolonged droughts over the grain belts of the world lead to shortages of food and in some cases widespread starvation. Dry spells existing over several years can deplete the water supplies of large cities and of hydroelectric operations. Extremely cold winters mean heavy fuel demands that sometimes exceed available supplies leading to cold homes and interrupted industry and commerce. Hot, humid summers strain the electrical capacities of most regions because of the widespread use of air conditioning, particularly in modern skyscrapers.

As everyone knows, heavy concentrations of air pollution or the occurrence of violent storms can lead to tragedy. Every year thunderstorms, lightning, tornadoes, and hurricanes account for huge losses of property and, more important, they injure and kill large numbers of people. At the same time, thunderstorms and hurricanes bring rainfall to nourish farm lands and fill reservoirs.

Clearly, the atmosphere affects the lives of everyone. We must know more about it. What are its properties? What causes the wind, the clouds, rain, snow, hail, and lightning? How can we more accurately predict the weather? Can we change or control it? There are a great many other questions that you could ask. The aim of this book is to supply the answers where they exist and to note those problems still awaiting solution.

A BRIEF LOOK
AT THE REST OF THE BOOK

Composition of the Atmosphere

Chapter 2 deals with the composition of the earth's atmosphere. The substance we call air is actually a mixture of gases. Nitrogen and oxygen constitute most of the atmosphere, but there are a great many other gases in small but important amounts. Particles of dust, smoke, and mists, which in combination with air are called aerosols, are commonly present in the atmosphere. When particles are abundant, the sky loses its blue color and distant objects become indistinct or disappear. Certain types of atmospheric particles play crucial roles in the formation of clouds, rain, and snow.

The concentration of atmospheric contaminants depends on the quantity emitted into the atmosphere, the chemical and physical processes that affect them, and the volume of air through which they are mixed. In order to estimate this last factor it is necessary to know the structure of the atmosphere, i.e., the variations of such properties as temperature, pressure, and wind.

Energy Transfer

Chapter 3 discusses the nature of energy transfer with particular emphasis on radiation. It is known that the energy to drive the winds and generate storms comes from the sun in the form of shortwave radiation, most of it being visible light. At the same time, the land, water, and air emit long-wave radiation, much of it escaping, and carrying heat, to outer space. Since the average temperature of the earth as a whole remains essentially constant, the incoming solar energy and outgoing terrestrial radiation must be essentially in balance. Most solar energy falls over the equatorial regions. It is transported toward the poles by atmospheric and oceanic currents. If this did not occur, there would be progressively higher temperatures at low latitudes and progressively lower ones at higher latitudes.

The Winds

The atmosphere is a restless medium. The motions of the air range from barely discernible, feathery breezes to the turbulent blasts of a tornado or hurricane.

As shown in Chapter 4, in order to account for the winds, it is necessary to recognize that the air is a fluid responding to the physical laws of motion. As specified by Newton's second law of motion, force equals mass times acceleration. If you can specify the forces at work, you can calculate the acceleration of the air and hence its velocity. The principal wind-driving forces arise because of variations of atmospheric pressure from place to place. The pressure forces serve to accelerate air from regions of high to regions of low pressure. Once the air begins to move, frictional forces act to slow it down. Finally, because the air moves over a rotating, nearly spherical earth, the air is subject to a deflection effect known as a Coriolis force. As a result of the various forces, in the Northern Hemisphere air tends to move counterclockwise around centers of low pressure and clockwise around centers of high pressure. For example, in a hurricane approaching the United States, the winds blow counterclockwise around the eye of the storm.

Vertical Air Motions

The air not only moves horizontally; it rises and sinks, and in so doing, it affects the state of the sky. Descending air has the effect of inhibiting cloud formation, while ascending air, if it rises enough, causes clouds. Sometimes there are weak upward air motions over large regions and as a result cloud layers cover the entire sky. At other times, particularly in the summer, violent updrafts and downdrafts accompany the formation of thunderstorms. The variable air motions over small distances account for the turbulence experienced by airplanes flying through thunderstorms.

Strong vertical air motions usually occur because warm air rises and cool air sinks, a type of air circulation called convection. When the temperature decreases markedly with height, the atmosphere is unstable. In such a circumstance strong convection currents can develop, particularly if the air is humid.

The rate at which air pollution accumulates depends on the degree of convection near the ground. When the atmosphere is stable, with little vertical air motion, pollutants are trapped in a shallow layer of air. As a result there can be a build up of contaminants and heavily polluted air.

Winds over the Globe

Because of the distinctive features of the earth, the configuration of land and seas, its atmosphere and its rotation, the average patterns of high and low pressure and of the wind currents are unique characteristics of this planet. Chapter 6 notes that among the well-known air streams are the trade winds at low latitudes and the westerly winds at middle latitudes.

The term general circulation is used when referring to the average air motions over the entire earth between the ground and the upper reaches of the atmosphere. In considering the general circulation, the atmosphere can be regarded as a giant heat engine. Radiant energy from the sun, absorbed mostly in the equatorial regions, is the source of power to drive the engine, i.e., the wind currents. By means of a set of equations of fluid motion, and assuming that energy and mass are conserved, it is possible to construct mathematical models of the general circulation of the atmosphere.

Air Masses and Fronts

In order to describe and forecast the weather, it is important to understand the concepts of air masses and fronts. As noted in Chapter 7, huge bodies of air, covering regions thousands of kilometers in diameter, are found to have fairly uniform thermal and humidity properties. For example, air masses originating over the tropical Atlantic and the Gulf of Mexico are warm and moist. When such an air mass encounters a cold, dry one moving southward from Canada, the two bodies of air do not mix readily. Instead, the cold, heavier air slides under the warm, lighter air and a transition zone develops between the two air masses. This zone is called a front. When humid air moves up a sloping front, clouds and precipitation are the usual result.

Frontal zones are favorable locations for the development of cyclones. This term, often used in meteorology, refers to centers of low pressure around which the winds blow. Cyclones are common features on weather maps. Most cyclones, particularly those associated with fronts, produce widespread clouds, rain, or snow. Cyclones frequently observed over hot, dry deserts in the summer are often free of clouds.

Clouds and Precipitation

The processes of cloud formation are described in Chapter 8. Everyone knows, from looking at the sky, that there are many types of clouds. Some are associated with fair weather, others yield rain, snow, or hail. Over the years, a number of cloud classifications have been developed. The one used by the weather services around the world depends mostly on the appearance of the clouds.

It is well known that most clouds are made up of huge numbers of minute water droplets formed as water molecules condense on particles in the atmosphere. High clouds are made up of small ice crystals.

Knowing cloud types and properties, it is sometimes possible to anticipate the arrival of fair or stormy weather. In some circumstances, for example, during the approach of a well-developed cyclone, there is a regular sequence of clouds signaling that a storm is on its way.

Contrary to the belief of many people, most raindrops are not grown by condensation. Instead, they are produced either by the collision and merging of cloud drops or by the melting of snowflakes or ice pellets. Snowflakes are aggregates of ice crystals that have collided and stuck together. Hail represents a spectacular class of frozen precipitation resulting when ice particles remain in the upper part of a thunderstorm long enough to accrete large amounts of ice. On rare occasions hailstones may be the size of oranges, but most damage is done by heavy downfalls of hail less than a centimeter in diameter.

The quantity of water substance on the earth is essentially constant. Water evaporates from oceans, rivers, and lakes and transpires from humans, animals, and plants. In the atmosphere, water vapor is converted to rain and snow that carry the water back to the surface of the earth. The movement of water through the geophysical system is called the hydrologic cycle.

Severe Storms

Thunderstorms range in intensity from those producing only a single flash of lightning and no rain at the ground to those yielding a great deal of lightning, torrential rains, damaging hail, and deadly tornadoes. These storms, discussed in Chapter 9, are dramatic examples of convection at work. When the atmosphere is moist and unstable, strong thunderstorm updrafts can extend to altitudes exceeding 15 km.

Although Florida experiences more thunderstorms than any other state, the Great Plains region of the United States is the most prolific in the world for the production of hail and tornadoes. They are most likely to occur when a shallow layer of warm, humid air from the Gulf of Mexico is surmounted by a layer of fast moving, dry air from the west.

Tornadoes occur frequently over the area extending from Texas through Oklahoma to Illinois and Indiana. They also are observed in other countries but not with the high frequency or intensity experienced in North America.

Hurricanes are much larger and longer lasting than tornadoes and as a consequence do more damage. Hurricanes occur over many tropical oceans, particularly in the Northern Hemisphere. In the western Pacific these violent tropical storms are called typhoons. In southeast Asia they are called cyclones. As hurricanes approach land, they cause a rise in ocean level and a surge of seawater over low-lying coastal areas. This destructive flooding is augmented by heavy rain and strong winds.

Meteorological satellites and radar are widely used for observing the development and movement of severe storms. Improvements of these detection techniques are leading to better forecasts and the saving of a large number of lives.

Sights and Sounds

The distinctive features of a thunderstorm are the lightning and thunder that define its existence. There are many other acoustical and optical phenomena in the atmosphere that are the source of wonder and pleasure. The roaring sound of an approaching tornado signals that disaster is on the way. On the other hand, a rainbow or a halo can be sights of unmatched beauty. Mirages produce images that sometimes defy the imagination. These interesting subjects are examined in Chapter 10.

Climate

In recent years there has been a resurgence of interest in the study of climate, which, for most purposes, can be defined as the average state of the weather elements of a place or region. The tabulation of weather reports has been going on for a long time, for more than 200 years in some localities. These data have been used to calculate the climatological characteristics of many places. As you would expect, there is an important relationship between climate and native vegetation. For this reason, vegetation types have been used by various climatologists in constructing climatic classifications of the world.

Ingenious methods are being used in the study of the climate of the entire earth. From geological evidence it is clear that over the several-billion-year history of this planet, temperatures have changed markedly. There have been ice ages interspersed with relatively warm interglacial periods. The patterns of cool and warm eras of various lengths have been found to continue up to the present time.

Climatic variations over the last hundred years are of particular interest. From about 1880 to 1940 there was a general increase of average air temperatures near the earth's surface, followed by a cooling trend. Recent evidence shows a leveling off of global temperatures. It still is not clear how to account for this sequence of events. Have human activities such as the emission into the atmosphere of gases and particles been contributing significantly to climate fluctuations on a global or regional scale? This question is addressed in Chapter 11.

Applications

Chapter 12 deals with the many ways that a knowldege of weather and climate can be used to benefit people. The design of airports, the industrial zoning of cities, the planning of agriculture are just a few functions that should take into account the statistical properties of temperature, rainfall, and wind.

Accurate weather forecasts, made as long in advance as possible, have great value for almost everyone, but particularly for those engaged in weather-sensitive occupations—farmers, builders, operators of motor vehicles and airplanes, just to name a few.

Accurate predictions of violent storms, particularly tornadoes, hurricanes, and those producing flash floods, can lead to the saving of lives even in cases in which little or nothing can be done to prevent property damage.

Recent advances in observational techniques have made it possible to obtain a much better knowledge of the temperature, pressure, and wind distribution in the atmosphere at any instant. Having such information, mathematical models of the atmosphere can be solved by means of electronic computers in order to calculate future states of the atmosphere. These numerical techniques have converted weather forecasting from an art to a science. The quality of forecasts has been improving over the last few decades and continued improvements are expected in the years to come.

Ultimately, it may be possible to control, or at least modify, the weather and climate. It has been speculated that one day we will be living in covered cities akin to huge, scaled-up versions of the Houston Astrodome. The practical difficulties of such proposals are enormous; if they come into being at all, it will be in the very distant future.

In the meantime, atmospheric scientists have been engaged in attempts to change the behavior of clouds and fog and to increase rain and snow. There have been efforts to reduce the fall of damaging hail and the intensity of hurricanes. Some progress has been made, but much more needs to be learned.

Appendices and Glossary

The appendices include tabulations of factors for converting units of measurements from the English system to the metric system. Students who are not familiar with metric units may wish to consult this section.

The last appendix lists sources of additional information for readers who wish to learn more about meteorology than is covered in this book. Teachers might want to use some of the listed educational films as supplements to the available reading material. The films are particularly helpful in illustrating physical processes at work in the laboratory and in showing atmospheric phenomena in motion.

Finally, at the end of the book is a glossary of technical terms commonly used in meteorology.

CHAPTER 2

Atmospheric Composition and Structure

THE EARTH'S ATMOSPHERE is a mixture of gases and *aerosols*, the latter being the name given to a system comprised of small liquid and solid particles distributed in the air. Aerosols will be discussed later; at this point we shall consider only the gases. As noted in the first chapter, air is not a specific gas. It is a mixture of many gases. Some of them may be regarded as permanent atmospheric components that remain in fixed proportions to the total gas volume. Other constituents vary in quantity from place to place and from time to time.

When the air is dry, that is, when there is no water vapor in it, the relative concentrations of various gases in the atmosphere are as shown in the upper part of Table 2-1. These quantities are essentially constant all over the earth and do not change up to an altitude of about 80 km. This deep layer through which the gaseous composition of the atmosphere is generally uniform is called the *homosphere*. At higher levels, as will be seen in a later section of this chapter, the chemical constituents of air change significantly with height. This layer is known as the *heterosphere*.

The principal sources of nitrogen, the most abundant constituent of air, are decaying agricultural debris and animal matter and the exhalations from volcanic eruptions. In addition, certain rocks release nitrogen, as does the burning of certain fuels. On the other side of the ledger, nitrogen is removed from the atmosphere by biological processes involving plants and sea life. To a lesser extent, lightning and high-temperature combustion processes convert nitrogen gas to nitrogen compounds that are washed out of the atmosphere by rain or snow. The destruction of nitrogen gas in the atmosphere is in balance with production, and the quantity is constant.

Oxygen, the second most abundant gas, is obviously crucial to life on earth. It is produced by vegetation which, in the photosynthetic growth process, takes up carbon dioxide and releases oxygen. It is removed from the atmosphere by man and animals whose respiratory systems are just the reverse of those of the plant communities. We inhale oxygen and exhale carbon dioxide. Oxygen dissolves in the lakes, rivers, and oceans where it serves to maintain marine organisms. It also is consumed in the process of decay of organic matter and in chemical reactions with many other substances. For example, the rusting of steel involves its oxidation.

Table 2-1 Principal Gases Composing the Earth's Atmosphere

Constituent	Percent by volume of dry air	Concentration in parts per million (ppm) of air
Nitrogen (N_2)	78.084	
Oxygen (O_2)	20.946	
Argon (A)	0.934	
Neon (Ne)	0.00182	18.2
Helium (He)	0.000524	5.24
Methane (CH_4)	0.00015	1.5
Krypton (Kr)	0.000114	1.14
Hydrogen (H_2)	0.00005	0.5
Important variable gases		
Water vapor (H_2O)	0–3	
Carbon dioxide (CO_2)*	0.0332	332
Carbon monoxide (CO)		<100
Sulfur dioxide (SO_2)		0–1
Nitrogen dioxide (NO_2)		0–0.2
Ozone (O_3)		0–2

*Carbon dioxide is uniformly mixed through the atmosphere, but it is increasing at a rate of about 0.7 ppm per year. Its concentration was about 332 ppm in 1978.

Not too many years ago it was suggested that, because of the widespread cutting of trees and reduction of plant life as well as other man-made environmental changes, there was a long-term threat to the supply of atmospheric oxygen. These fears were put to rest by an examination of measurements of oxygen concentration over this century. They showed no measurable change of oxygen concentration, indicating that production is in balance with destruction.

From the human point of view, the scarce, highly variable gases listed in Table 2-1 are of great importance. Water vapor is the source of rain and snow without which we could not survive. From common experiences it is well known that the water vapor content of air varies a great deal. In a desert region the quantity of water vapor can be so low as to represent only a tiny fraction of the air volume. At the other extreme, in hot, moist air near sea level, say over an equatorial ocean, water vapor may account for as much as 3 percent of the air volume.

When a meteorologist talks about "moist air," he means air containing a great deal of *water vapor*, that is H_2O in a gaseous phase. If clouds exist, the atmosphere also contains H_2O in the liquid phase in the form of tiny water droplets.

There are many ways to express the degree of moisture in the atmosphere. More will be said about this subject later, but it is important to recognize that the most common measure, the *relative humidity*, as the word relative indicates, is not an absolute measure of the quantity of water vapor. The relative humidity is the *ratio*, expressed in percent, of the quantity of vapor in the air to the maximum amount the air can contain. The latter decreases as the temperature decreases. Therefore, as the temperature decreases, the relative humidity increases even when the mass of water vapor in a volume of air remains unchanged. This accounts for the fact that on most days the relative humidity is highest during the early morning hours when air temperature has its minimum value.

On some days early morning temperatures fall so low that the relative humidity approaches 100 percent, and as a result dewdrops form on blades of grass and other objects. With the rise of the sun and increasing temperatures, the relative humidity diminishes and the dew evaporates.

From place to place and from time to time there are large variations of water vapor in the atmosphere, but the total quantity over the entire earth is virtually constant. The same cannot be said about carbon dioxide (CO_2). The concentration of this sparse but important gas has been increasing for the last hundred years or so (see Fig. 2-1). In 1977 it averaged about 332 parts per million* of air.

There are various sources of atmospheric carbon dioxide (CO_2); they are the decay of plant material and humus in the soil, and the burning of fossil fuels—coal, oil, and gas. The sinks of CO_2 are the oceans and plant life that uses CO_2 in photosynthesis. In 1978 atmospheric chemists were still debating whether or not the biosphere is a net source or net sink of CO_2. Some claim that the harvesting and clearing of forests has led to a net increase of atmospheric CO_2. Others have concluded that photosynthetic use of CO_2 exceeds the amount released by biological processes. There is agreement that the oceans represent the major sink, and recent evidence indicates that the fraction going into the seas is diminishing. During the 1970's atmospheric CO_2 was accumulating at a rate of about 0.7 ppm per year but, as shown in Fig. 2-1, it is expected to increase much more rapidly in decades to come. It is expected that the fraction of carbon dioxide going into the oceans will continue to diminish while the total amount released will increase at least through the end of this century.

The intense interest in atmospheric carbon dioxide is not based on any concern about the toxic effects of this gas. Instead, the attention comes from the fact the carbon dioxide is a good absorber of infrared radiation, that radiation emitted by the earth and the atmosphere. This topic will be discussed in detail in the next chapter. At this point it is relevant to recognize that such absorption acts to warm up the lower atmosphere. It has been

*The concentration of small amounts of a substance are often measured in parts per million (ppm), a unit used to express fractional quantities of one substance in relation to the quantity of another substance. The concentration of 332 ppm by volume means that there are 332 parts of carbon dioxide in a million parts of air. It corresponds to 0.0332 percent of the total volume of dry air.

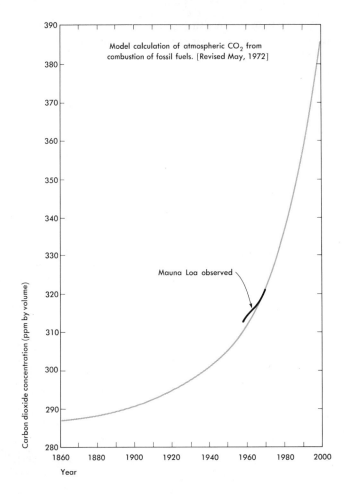

Figure 2-1   Calculations of carbon dioxide concentrations in the earth's atmosphere showing increases attributed to the combustion of fossil fuels. *From* L. Machta, Brookhaven Symposium, May 1972.

proposed that worldwide carbon dioxide increases can cause changes in the global climate. Some scientists have proposed that such alterations have already begun. This subject is a complicated one that will be dealt with later.

Ozone (O_3), another important, highly variable gas, occurs mostly at upper altitudes, but it is also found in urban localities having a great deal of industry and automotive traffic and a generous supply of sunshine. In cities such as Los Angeles ozone concentration may be more than 0.1 ppm in extreme cases.

Most atmospheric ozone, as shown in Fig. 2-2, occurs at altitudes from 10 km to 50 km above the ground. Between 15 km and 30 km ozone concentrations often exceed 1.0 ppm and may be as large as 10 ppm. The concentrations vary greatly with latitude, season, time of day, and weather patterns. The ozone layer is maintained by a large number of photochemical reactions. In general, the following important processes occur: Molecules of oxygen gas (O_2) absorb ultraviolet solar radiation, and as a result the molecules are dissociated to form oxygen atoms (O). Collisions of oxygen molecules (O_2) and oxygen atoms (O) and other particles lead to the formation of ozone (O_3). They in turn absorb ultraviolet radiation and are dissociated to form oxygen molecules and oxygen atoms. In actuality, there are many reactions causing ozone formation and destruction. The net result is a layer of ozone, which in cross section resembles the one shown in Fig. 2-2.

The ozone layer is important because by absorbing ultraviolet radiation in the upper atmosphere, it reduces the amount reaching the surface of the earth. Exposure to increased doses of ultraviolet rays would cause more severe sunburns and would increase the risk of skin cancers. Agricultural specialists indicate that a substantial increase in ultraviolet radiation could also affect other components of the biosphere.

Interest in the ozone layer has surged since 1970 when the United States government was considering major new investments in the design of a supersonic transport airplane (SST). It was anticipated that it would fly at about three times the speed of sound at a cruising altitude of about 20 km.

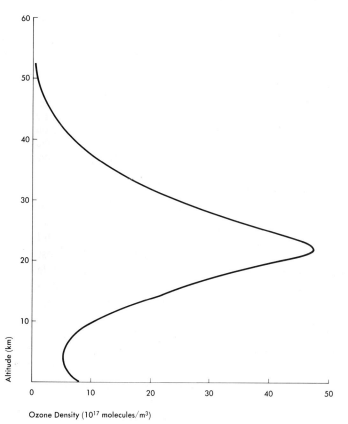

Figure 2-2 Average middle latitude ozone number concentration proportional to ozone density. *Data from* A. J. Krueger and R. A. Minzner, NASA TM X-651-73-22, 1973.

Altitude (km)

Ozone Density (10^{17} molecules/m^3)

A glance at Fig. 2-2 shows that such flights would be in the ozone layer. Certain scientists opposed the development of SST airplanes on the ground that nitrogen oxide gases emitted by the engines would react with and reduce ozone amounts. These early speculations were subsequently reaffirmed in 1975 in a report by the U.S. National Academy of Sciences. It concluded that unless the nitrogen oxides from the SST engines could be greatly reduced, large fleets of airplanes posed a serious threat to the ozone layer.

The SST controversy focused attention on the ozone layer and, as a consequence, a number of other important discoveries were made. In the early 1970's various scientists noted that the nitrogen in certain heavily used fertilizers might, over a long period of time, also represent a threat to ozone in the atmosphere. It was also found that, on a shorter time scale, the explosions of nuclear weapons in the atmosphere can inject massive quantities of nitrogen oxides into the ozone layer. Such an event could cause a marked reduction of ozone amounts. It has been estimated by a prominent group of scientists that an all-out nuclear war, as staggering as such a conception is to contemplate, would have as a terrible byproduct the temporary destruction of from 30 percent to 70 percent of the earth's ozone.

In the middle 1970's many people became concerned about the effects of fluorocarbons on the ozone layer. These substances, which are manufactured under the trademark of Freon, have been used in aerosol spray cans and refrigerating systems. Freon compounds are highly volatile, but they are not flammable or toxic, do not react with other components, and are relatively insoluble in water. After they are released into the air, the inert molecules diffuse slowly into the upper atmosphere, taking several years to reach the upper part of the ozone layer at altitudes of from 25 km to 35 km. At these elevations ultraviolet radiation from the sun causes the dissociation of fluorocarbon molecules and the release of chlorine atoms. They react with and destroy ozone molecules.

On the basis of available knowledge about rates of release of fluorocarbon gases into the atmosphere, the rates of diffusion to the upper atmosphere, and the speed of photochemical reactions in the upper atmosphere, calculations have been made of the rates of depletion of ozone. Although there are still some uncertainties, there is widespread agreement that continued large releases of fluorocarbons on a worldwide scale are likely to cause significant reductions of stratospheric ozone. The maximum effects are not likely to be observed until many years into the future—after 1990. It has been further predicted that the reduction of ozone quantities would lead to significant increases in skin cancer in humans. It has also been suggested that since fluorocarbons absorb radiation from the earth, and inhibit its escape to outer space, increases of fluorocarbons in the atmosphere might, over the next century, cause a warming of the earth. In 1977 the United States government adopted regulations intended to greatly reduce the use of fluorocarbons in aerosol spray cans.

Because of the slow processes involved, it may be many years before it is known for sure if fluorocarbons pose the serious threats that have been predicted. A few scientists have argued that the ozone layer in the upper atmosphere has evolved over the ages and is not a fragile component of the earth's atmosphere. Further research should help to resolve the different views on this important subject.

GASEOUS POLLUTANTS IN THE ATMOSPHERE

As everyone knows, certain gases, if they exist in sufficiently high concentrations, can be toxic to people, animals, and plant life. For example, when ozone occurs in high concentrations, it is toxic to biological organisms. This does not happen often, but in heavily polluted localities, such as Los Angeles, ozone near the ground sometimes is sufficiently abundant to cause leaf damage to certain plant species.

Very large quantities of potentially hazardous gases are introduced into the atmosphere as a result of human activities. Air pollutants are emitted from furnaces, factories, refineries, smelters, and engines, particularly automobile engines. All these things and others like them burn fossil fuels—coal, oil, gasoline, and kerosene. In the process they emit gases and smoke particles which may spend a great deal of time in the atmosphere reacting with other substances and causing the formation of toxic compounds.

The most widespread and potentially hazardous gaseous pollutants are carbon monoxide, sulfur dioxide, nitrogen oxides, and hydrocarbons. The last of these compounds comes from vaporized gasoline and other petroleum products. At every service station some gasoline evaporates during the tank-filling process. More importantly, gasoline fumes pass out of the tailpipe of most cars because the engine fails to burn the gasoline completely. This problem was more serious before about 1970 because there were virtually no controls over the emissions of automobile engines. With the enactment of emission standards by the federal government, motor vehicle manufacturers have added control devices to motor vehicles in order to improve engine efficiency and reduce hydrocarbon emissions.

When hydrocarbons get into the air on a sunny day, they react with nitrogen oxide molecules, also emitted by automobile engines. The results of the photochemical reactions are a series of complicated organic molecules called *peroxyacyl nitrates* or PAN for short. Under certain conditions, when the air is humid, particles composed of PAN can be in sufficient number to produce a hazy condition known as *smog*. In concentrations of a few parts per million, these particles can cause the eye irritation well known in southern California on a polluted day. Fortunately, the eye irritation is a temporary condition and does not lead to serious eye damage. There is some concern, however, that in extreme instances photochemical smog particles may contribute to lung ailments.

Nitrogen oxides are produced by the combination of nitrogen and oxygen in a combustion process carried out at high temperatures. The cylinders of an automobile or the engines of airplanes provide excellent conditions for such chemical reactions. Earlier it was noted that nitrogen oxides released at high levels may react with ozone and reduce its concentration with attendant increases of ultraviolet radiation reaching the earth.

Automobile engines are ground-level sources of nitrogen dioxide (NO_2), a gas which can be toxic in sufficiently high concentrations. As already noted, in the atmosphere nitrogen dioxide reacts with hydrocarbons and is converted to photochemical smog. It has been suggested that if hydrocarbon emission is reduced, there will be an increase in nitrogen dioxide in the air. Fortunately, the concentrations of this pollutant in the atmosphere have been below the level where humans or animals are injured.

Another well-known air contaminant contributed largely by automobiles is carbon monoxide (CO). Unlike the much more abundant carbon dioxide (CO_2), it is toxic when concentrations exceed about 10 ppm. Every year people die as a result of running an automobile engine in a closed garage or because carbon monoxide leaks into cars from defective exhaust systems. Carbon monoxide attacks the hemoglobin in the blood and prevents it from transporting oxygen from the lungs to the tissue of the body. Fortunately, in the open air the concentration usually is too low to have direct effects on human health. There is evidence, however, that in large cities, under stagnant air conditions, carbon monoxide concentrations may be high enough to cause dizziness and headaches. Exposures to heavy doses of carbon monoxide have been implicated as partly responsible for some automobile accidents. Incidentally, cigarette smoking causes the ingestion of large concentrations of carbon monoxide.

Sulfur dioxide is a gaseous pollutant regarded as a serious threat to plants, animals, and people as well as to many nonorganic substances such as limestone. Most of this gas is introduced into the atmosphere by natural processes—volcanic eruptions, sea spray, and decay of organic matter. In addition, large amounts come from the sulfur contained in coal and oil and from certain smelting operations.

In the presence of water vapor, sulfur dioxide, through a series of chemical reactions, is converted to droplets of sulfuric acid (H_2SO_4). These droplets may combine with smoke particles in the air, yielding a product that can have important effects on many natural processes. The acid bearing particles can be breathed into the lungs of humans and animals. They can be deposited on plants and on other objects. The old churches and statuary of Europe exposed to the slow but steady attack by airborne sulfuric-acid particles are becoming seriously crumbled and disfigured.

The sulfur-bearing particles in the atmosphere serve as nuclei for the formation of cloud droplets and ultimately are carried to the ground by rain and snow. Over the last few decades there has been growing evidence that rain has been more acidic downwind of major industrial areas. The sulfur compounds contained in rain and snow serve at least one important beneficial function, because sulfur is a necessary soil ingredient for the growth of various crops. When natural processes fail to supply adequate amounts, farmers use fertilizers to meet the soil's needs.

On the other hand, downwind of large industrial regions releasing large quantities of sulfer dioxide, the concentrations sometimes are sufficiently high to do major damage to agriculture. In extreme cases in past years there have been examples of such high sulfur dioxide concentrations as to cause the soil to be so highly acidified that nothing would grow. Fortunately, in the United States, the adoption of environmental control procedures is reducing the emission of sulfur into the atmosphere.

PARTICLES
IN THE ATMOSPHERE

The atmosphere contains huge numbers of solid and liquid particles. The largest ones, which constitute clouds, rain, snow, and hail, will be discussed in a later chapter. At this time we consider only the small ones most of which are invisible to the unaided eye. They are composed of the following: soil blown into the air; salts remaining when droplets of ocean water dry out; smoke from combustion processes; various substances thrown into the atmosphere by volcanoes; sulfate and nitrate particles produced by chemical processes in the atmosphere; and tiny droplets of sulfuric or nitric acid formed in the air.

Certain particles in the air that play crucial roles in the development of cloud droplets and ice crystals are discussed in Chapter 8.

Many atmospheric particles are regarded as pollutants. In sufficient quantities and sizes they give the sky a milky appearance as they diffuse sunlight. When the concentration of particles is sufficiently great, particularly when humidities are high, the result can be a marked reduction in visibility (Fig. 2-3). In severe cases, it may be reduced to less than a hundred meters and lead to interruptions in aircraft traffic.

Atmospheric particles in smog, formed by the action of chemical processes involving hydrocarbons and sunlight, are a source of eye irritation. Potentially much more critical is the role of particles in causing lung problems on days when the air is heavily polluted. It is suspected that particles can carry minute but significant quantities of sulfuric acid. With every breath, large numbers of particles are introduced into the lungs and some of them

Figure 2-3 Photograph of New York City taken from the eightieth floor of the Empire State Building during an air pollution alert on October 28, 1966. *Courtesy* P. C. Freudenthal, Consolidated Edison of New York, Inc.

are deposited on the respiratory tract. It is important to recognize that the likelihood of particles being deposited depends on their sizes. Those only a few tenths of a micrometer* in radius tend to avoid capture as they move in and out of the respiratory system. Most of the particles having radii greater than a couple of micrometers never make it into the lungs because they are captured in the nose and throat. As shown in Fig. 2-4, the particles most effectively deposited in the respiratory tract are those close to one 1 μm in radius. They are the ones that are the source of danger to your health.

There are many ways to observe particles in the air. They can be captured on slides and viewed by means of an optical or electron microscope.

*A micrometer (μm) is a millionth of a meter. A typical human hair has a radius of about 50 μm.

20

In some instances, chemical tests can be used to identify their chemical compositions. The concentration of particles can be measured with an Aitken nuclei counter. This device consists of a chamber that takes a sample of air and expands it rapidly. As a result of the expansion, the air is cooled suddenly and, for reasons to be discussed in Chapter 8, water vapor condenses on the particles and a cloud is formed. By measuring the opacity of the resulting cloud, it is possible to estimate the concentration of particles that were present in the air and had radii greater than about 0.001 μm.

An estimate of the particulate loading, that is, the mass of particles in a unit volume of air, can be obtained by means of a filter through which a large volume of air is passed. The filter is weighed before and after the sample is taken to determine the mass of particles. Certain chemical tests allow evaluation of the relative amounts of particular substances. Unfortunately this procedure gives almost no information about the sizes of the particles.

The overall particulate loading of the atmosphere can also be estimated by optical techniques. For example, measurements of the changes in incoming solar radiation on cloudless days can be related to the transparency of the atmosphere. As the quantities of smoke, dust, and haze are increased, the transparency is reduced. Meteorologists say that this indicates an increase in turbidity. The reflection of light from an intense searchlight beam, or better still from a laser beam pointing upward at night, can be used to locate layers of particles and estimate their sizes and concentrations.

Table 2-2 presents some data on the concentrations of particles observed in the atmosphere by means of Aitken nuclei counters. The table shows that

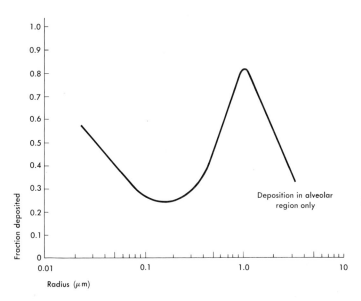

Figure 2-4 Calculated fraction of particles deposited in human lungs as a function of particle radius plotted logarithmically. *From* Air Quality Criteria for Particulate Matter, National Air Pollution Control Administration, AP-49, 1969.

over continents, particularly in urban areas, the particle concentrations are very much higher than over the oceans.

In the atmosphere the smaller the particles, the greater the number.* Particles having radii of about 0.1 μm may occur in concentrations of about 10^4 per cubic centimeter while those of radii about 10 μm may be as few as 0.01 per cubic centimeter.

Table 2-2   Concentration of Aitken Particles in the Atmosphere

Location	Number of measurements	Concentrations (particles/cm³)		
		Average	*Minimum*	*Maximum*
Ocean	600	940	2	39,800
Mountain				
Above 2 km	190	950	6	27,000
1–2 km	1000	2130	0	37,000
0.5–1 km	870	6000	30	155,000
Country, inland	3500	9500	180	336,000
Town	4700	34,300	620	400,000
City	2500	147,000	3500	4,000,000

Source: H. Landsberg, Atmospheric condensation nuclei. *Ergebnisse der Kosmischen Physik*, 3, 1938.

The average residence time of particles in the atmosphere depends on where they are located. Those in the lower atmosphere remain in the air from only 1 to 4 weeks. Tiny particles thrown into the stratosphere, above about 10 km, by volcanic eruptions may stay there for a year or two. The absence of clouds and rain in the stratosphere and the small fall velocities of the particles account for these long periods.

It is commonly observed that over cities, particularly large ones, there tends to be a cloud of smoke and dust particles. A crucial question, still not satisfactorily answered, is the degree to which human activity is increasing the concentration of particulates in the whole atmosphere. Since they interfere with the transfer of radiant energy through the air, a significant increase in the particle content might have an important effect on the global climate (see Chapter 11).

The available evidence indicates that there has been an increase in particles over the Northern Hemisphere, particularly over the northern half of the oceans, but little change over the Southern Hemisphere. Figure 2-5 shows a curve of atmospheric turbidity based on data collected on a mountain top in Hawaii. The greater the turbidity, the greater the quantity of particulate

---

*For representing very large or small quatities, it is often convenient to use scientific notations in which powers of ten are used in place of many zeroes. For example, $1000 = 10^3$; $320,000,000 = 3.2 \times 10^8$. Also $1/1000 = 0.001 = 10^{-3}$; $1/10,000,000 = 0.0000001 = 10^{-7}$.

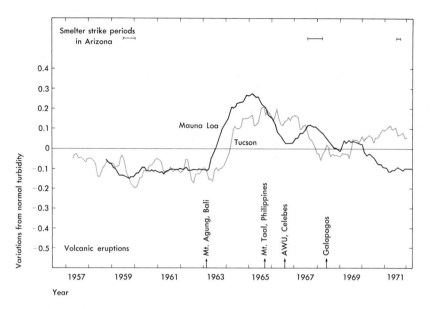

Figure 2-5 Departure from normal of atmospheric turbidity measured over Tucson, Arizona and Mauna Loa, Hawaii. The curves are based on yearly averages. *From* K. Heidel, *Science*, 1972 **177**: 882–883.

matter in the air. These observations clearly show the injection of particles into the atmosphere by several major volcanoes.

Following the massive eruptions of Krakatoa in 1883 and Mount Agung in 1963 sunsets were particularly brilliant and colorful for most of the succeeding two years. Minute particles scattered the blue colors of sunlight more than the red and led to magnificently colored skies at sunrise and sunset. Since 1963 there have been no really massive volcanic explosions, and the turbidity level has returned to close to its value of earlier years.

PRESSURE AND TEMPERATURE OF THE AIR

It is well known that in relation to the size of the earth, the atmosphere is extremely shallow. You might ask, in light of the gaseous nature of the atmosphere, how one defines the depth of the atmosphere. It can be done by specifying how the mass of the atmosphere varies with height and noting what fraction of the mass exists below any particular altitude. In order to do this it is necessary to know how the mass of a cubic meter of air varies with atmospheric temperature and pressure.

Everyone knows that temperature measures the degree of heat or cold of any substance. Temperature is a representation of the state of motion of the atoms or molecules making up the substance. The higher the temperature, the higher the energy of motion, i.e., the higher the kinetic energy of the atomic constituents. When a hot body is brought into contact with a cold one, heat is conducted from one to the other. The molecules composing the warm material are caused to move slower, while those composing the cold material are accelerated.

There are various scales for measuring temperature, the most common one in English-speaking countries being the *Fahrenheit scale*. Two reference points on this scale are 32°F and 212°F, the values at which water freezes and boils at sea level, respectively. In most of the world and among virtually all scientists, the *Celsius scale* is the rule. In this system, 0°C and 100°C are the freezing and boiling points of water. Since the number of degrees between freezing and boiling are 180°F and 100°C, each 1°F is equal to 100/180 or 5/9°C. Therefore, in order to convert from a temperature (F) in degrees Fahrenheit to (C), the corresponding temperature in degrees Celsius, one can use the formula

$$C = (F - 32) \times \frac{5}{9}$$

For example, if the temperature is 50°F, it equals $(50 - 32)5/9 = 10°C$.

A person has to use the Celsius scale for a while before the temperature levels are readily understood in terms of everyday activities. For example, not knowing the Celsius scale, a sick American traveler in France might be needlessly shocked to hear a doctor say that his or her temperature was 37 degrees. In order to put the Celsius scale in its proper perspective, it would be helpful to become familiar with the Celsius temperatures listed in Table 2-3.

The temperature scale that must be used in equations describing the physical state of a substance or physical and chemical processes is called the

Table 2-3 SOME CELSIUS TEMPERATURES WORTH REMEMBERING

Temperature		*Everyday association*
(°C)	*(°F)*	
−40	−40	An extremely cold day in the U.S.
0	32	Freezing point of water
10	50	A cool day
20	68	A cool room temperature
30	86	Swimming weather
40	104	An extremely hot day in the U.S.

Kelvin scale. It gives the temperature above *absolute zero,* the temperature at which all atomic and molecular motions stop. The term *degrees absolute* is taken to be synonymous with *degrees Kelvin.*

The temperature in degrees Kelvin (K) is given by

$$K = C + 273$$

Therefore, absolute zero is at $-273°C$. Also, the freezing and boiling points of water are at $273°K$ and $383°K$, respectively.

Although most people are familiar with examples of the effects of air pressure through experiences with bicycle or automobile tires, the quantitative meaning of the term *pressure* is sometimes misunderstood. Pressure is defined as a force per unit area. For example, if a force equal to F were uniformly applied on a surface of area A, the pressure exerted on the area would be F/A.

Meteorologists regularly record atmospheric pressure at the earth's surface and note how it changes with time. The pressure force is actually the weight of the air in a column extending from the surface to the top of the atmosphere. On the average, *at sea level,* the air in a column having a cross-sectional area of 1 sq. in. weighs 14.7 lb. This may be a surprisingly large value to people accustomed to thinking of air as an unsubstantial medium.

There are many units for the measurement of pressure. The interested reader is referred to Appendix I, but it is necessary to mention that meteorologists usually express pressure in *millibars.* The average sea level value is 1013.25 mb; for mathematical convenience, this is sometimes rounded off to 1000 mb.*

Atmospheric pressure is measured by means of an instrument called a *barometer.* The most common barometer observes the height of a mercury column supported by atmospheric pressure. A simple version of the mercury barometer can be made in the following way. A glass tube about 1 m in length and closed at one end is filled with mercury. The open end is sealed and the tube is tipped over and placed in a small reservoir of mercury. When the seal is removed from the open end of the tube, some of the mercury flows out, but most of it remains in the tube. The force of atmospheric pressure on the surface of the mercury holds the mercury column in suspension. This is illustrated in Fig. 2-6. Average sea-level pressure supports a column of mercury about 760 mm or 29.92 in. long. The length of the column decreases or increases as atmospheric pressure decreases or increases, respectively.

There are many other ways to observe atmospheric pressure. Instruments that record variations of pressure with time, called *barographs,* usually use an aneroid barometer (Fig. 2-7). In essence, it is a shallow, partially evacuated container that can expand and contract as outside pressure

*In physical units a millibar is equal to 100 N/m^2, where a newton is the force which must be exerted on a mass of 1 kg in order to accelerate it at a rate of 1 m/s^2.

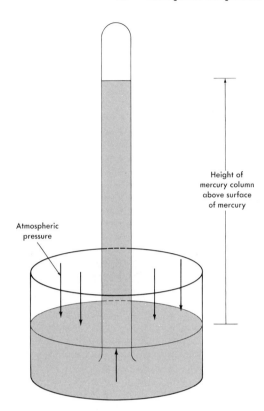

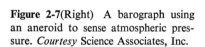

Figure 2-6(Left)   The height of the mercury columns is a measure of atmospheric pressure. The weight of the mercury is just balanced by the upward pressure force. In a fluid a pressure force acts equally in all directions.

Figure 2-7(Right)   A barograph using an aneroid to sense atmospheric pressure. *Courtesy* Science Associates, Inc.

decreases or increases. Aneroids are used widely in aircraft altimeters, devices for observing aircraft altitudes. Since atmospheric pressure decreases with height, a pressure-sensing instrument can be calibrated in terms of height.

The variation of pressure with height can be calculated if one knows the *density*, which is defined as the mass of a substance in a unit volume. For example, one cubic centimeter of water has a mass of about one gram; therefore, the density of water is about 1 g/cm³. As indicated in Table 2-4, at typical sea-level conditions, air has a density of about 0.001 g/cm³.

Table 2-4　Density of Some Common Substances

	Density	
Substance	(g/cm³)	(kg/m³)
Air:　$p = 1000$ mb　$T =$　20°C	0.001188	1.188
$= 500$　　　$= -20$	0.000688	0.688
$= 100$　　　$= -50$	0.000156	0.156
Ice:　　　　　$T =$　0°C	0.917	917
Water:　　　　$T =$　0°C	0.99984	999.84
$=$　4	0.99997	999.97
$=$　8	0.99985	999.85
Dry soil	1.7*	1700*
Granite	2.7*	2700*
Mercury	13.595	13,595

*Approximate.

Table 2-4 shows that the density of pure water at 4°C is greater than its density at 0°C. As a matter of fact, water has its highest density at a temperature of 4°C. The more dense, warmer water sinks in relation to the less dense, colder water at 0°C. For this reason, as air temperatures drop below freezing over water bodies, the colder upper surface begins to freeze first. If temperatures are sufficiently low, a layer of ice forms and serves to insulate the remaining, warmer, unfrozen water. This allows fish and other living creatures to survive the winter.

The density of a material is a measure of the numbers and masses of atoms or molecules in a unit volume. If the distribution of constituents remains unchanged, as it nearly does in air up to an altitude of perhaps 80 km, the density varies with the number of gas molecules. On the average, the number of molecules per unit volume decreases with altitude and the air density does the same (see Table 2-5).

The density of a substance is sometimes said to measure how heavy it is, but this is not strictly correct. The term *heavy* usually refers to weight, and weight is a statement of force rather than of mass. Let us consider this point further by taking an example. The mass of a typical adult male might be 70 kg. This is the quantity of matter making up the individual and it would be the same whether he is walking on the earth or the moon or circling the earth in a satellite. The person's density would be 70 kg divided by the volume of the body; as is the case with mass, density is independent of where it is measured.

The *weight* of the person is given by the mass multiplied by the acceleration of gravity.* Weight represents the force of gravity pulling the person toward the center of the earth. For example, when you stand on a bathroom scale, it is depressed by the downward force you exert.

Because the moon has a mass about one-sixth that of the earth, the gravitational acceleration on the moon also is one-sixth that of the earth. As a result, a person's weight on the moon would be about one-sixth its value on the earth. In an orbiting satellite the earth's gravitational pull is very nearly balanced by the outward centrifugal force, the net acceleration is nearly zero, and, therefore, a person's weight is nearly zero. This accounts for the fact that large masses of men and equipment can be suspended in an orbiting satellite.

Air pressure can be related to density (mass per unit volume) and temperature through the equation of state for an ideal gas.† From measurements of pressure and temperature as a function of height, you can calculate the appropriate values of air density. Annual averages for 15° north latitude are given in Table 2-5. Before examining certain features of these data, it is in order to say a few words about techniques used by meteorologists for measuring pressure, temperature, and humidity in the atmosphere above the earth's surface.

The *radiosonde* is the standard instrument used all over the world for obtaining observations of atmospheric properties (see Fig. 2-8). It consists of an aneroid barometer as well as temperature-sensing and relative-humidity-sensing elements. The output of these elements are used to modulate the output of a small radio transmitter. The entire package is supported under a balloon whose size and weight can lift the radiosonde at a rate of about 5 m/s (1000 ft/min). As the balloon rises, it expands until it ultimately breaks and the radiosonde is lowered to the ground on a small parachute. As the device ascends slowly, generally to altitudes approaching 30 km, as well as during its more rapid descent, it transmits meteorological information back to a radio receiver.

---

*This can be written $W = mg$, where W and m are weight and mass, respectively, and g is the acceleration of gravity and equals 9.8 m/s².

†Atmospheric pressure can be calculated from the ideal gas law;

$$p = \frac{R}{m} \times T \times \rho$$

where p is pressure, T is absolute temperature, ρ is density, R is the universal gas constant, and m is the molecular weight of the gas. For air $m = 28.9$ and for water vapor $m = 18$. Therefore, for the same pressure and temperature, the density of water vapor is less than the density of air.

Table 2-5 ANNUAL AVERAGE TEMPERATURES, PRESSURES, AND DENSITIES AT 15° N LATITUDE

Altitude		Temperature	Pressure		Density*
(km)	(mi)	(°C)	(mb)	(% of 1013)	(kg/m³)
0	0.0	29.4	1013.0	100.0	1.17
5	3.1	− 2.5	559.0	55.0	7.20×10^{-1}
10	6.2	−36.2	285.0	28.0	4.20×10^{-1}
20	12.4	−66.4	56.0	6.0	9.52×10^{-2}
30	18.6	−40.9	12.2	1.0	1.83×10^{-2}
40	24.8	−19.2	3.0	0.3	4.18×10^{-3}
50	31.1	− 3.0	0.85	0.08	1.10×10^{-3}
60	37.3	−20.0	0.24	0.02	3.29×10^{-4}
70	43.5	−54.3	0.058	0.006	9.21×10^{-5}
80	49.7	−88.4	0.011	0.001	2.09×10^{-5}
90	55.9	−96.1	0.0017	0.0002	3.38×10^{-6}
100	62.2	−82.4	0.00029	0.00003	5.15×10^{-7}

*Standard international units for density are kilograms per cubic meter. In units of grams per cubic centimeter, densities are one thousandth of those listed in the table.

Source: U.S. Standard Atmosphere, U.S. Government Printing Office, Washington, D.C., 1966.

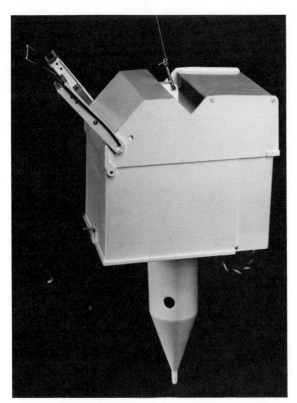

Figure 2-8 A radiosonde. The thin, white wire in the upper left-hand corner of the photograph is the temperature sensor. The downward projecting cone contains the antenna for telemetering data on temperature, humidity, and pressure. *Courtesy* Bendix, Environmental Science Division.

By means of directional antennas, radiosonde transmitters can be tracked as they rise and move with the wind. By noting the horizontal displacement as a function of time and height, it is a straightforward matter to calculate the winds along the trajectory of the balloon. When a radiosonde is used to obtain winds, the device is called a *rawinsonde*.

There are rawinsonde stations scattered over the globe, most of them over the land areas of the Northern Hemisphere; about 80 stations are located in the United States (see Fig. 2-9). According to international agreements, formulated by the World Meteorological Organization, upper air soundings are taken twice per day—at midnight and at noon Greenwich Mean Time (GMT). These times, written 0000 GMT and 1200 GMT, correspond to different local times, depending mostly on longitude. For example, in the midwestern United States, where Central Standard Time (CST) is used, rawinsonde observations are taken at 0600 CST and 1800 CST. The nearly simultaneous release of balloons at all stations makes it possible to construct *synoptic weather maps*, that is, maps showing conditions at any one time.

Observations of atmospheric properties at altitudes above about 30 km are made by means of specially instrumented sounding rockets and satellites. In the latter case, remote sensing instruments must be used. Special cameras record the clouds by day and night. Spectrometers are used to observe the character of the radiation from the atmosphere below the satellite. From the data it is possible to calculate the variations with height of temperature, humidity, and ozone. At this time, the calculated values are not as precise as atmospheric scientists would like them to be, but this is a relatively new and advancing area of research.

As noted earlier, atmospheric pressure is the weight of air over a unit area in a column extending to the top of the atmosphere. The weight depends on the density of the air in the column. As the air density through a column of air varies, the pressure at its base also varies. If the pressure at the ground and the height distribution of density is known, the pressure at any altitude can be calculated by means of a formula called the *hydrostatic equation*.* It states that the pressure difference between any altitude h_1 and a higher altitude

---

*The hydrostatic equation is $p_1 - p_2 = \rho g(h_2 - h_1)$ where ρ is the average density, g is the acceleration of gravity, and p_2 and p_1 are the pressures at height h_2 and h_1, respectively. For example, if p_1, the pressure at sea level, is 1013 mb, and if the average density, ρ, between the surface ($h_1 = 0$) and $h_2 = 1$ km is 1.1 kg/m³,

$$p_1 - p_2 = 1.1 \frac{\text{kg}}{\text{m}^3} \times 9.8 \frac{\text{m}}{\text{s}^2} \times 10^3 \text{ m}$$

$$= 10{,}780 \frac{\text{kg} \cdot \text{m}}{\text{m}^2 \cdot \text{s}^2} = 10{,}780 \frac{\text{N}}{\text{m}^2} = 108 \text{ mb}$$

since 1 Newton (N) $= 1 \frac{\text{kg} \cdot \text{m}}{\text{s}^2}$ and 1 mb = 100 N/m².

Therefore, p_2, the pressure at an altitude of 1 km, equals $(1013 - 108) = 905$ mb.

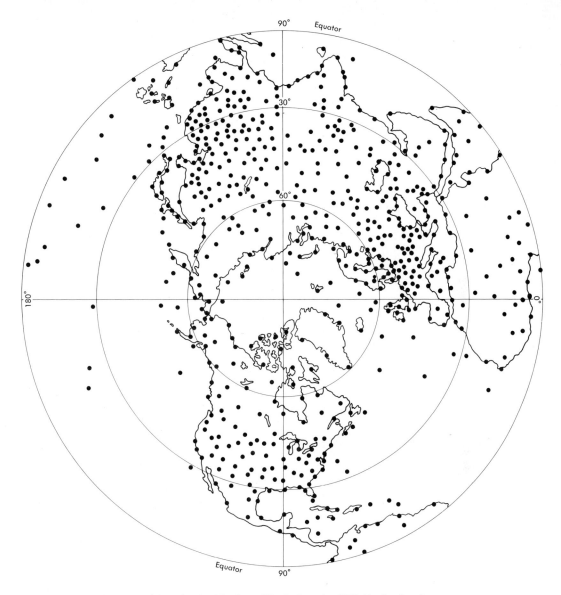

Figure 2-9 Radiosonde stations in the Northern Hemisphere in 1975. In the Southern Hemisphere there were about 146 stations, mostly at latitudes below 40°S. *Data from* World Meteorological Organization, Geneva.

h_2 is the weight of the air, in a column of unit cross-sectional area, between h_2 and h_1 (Fig. 2-10). For example, the unit area might be 1 m² or 1 in.². By taking a series of altitude layers, the pressure at succeedingly higher altitudes can be calculated step by step. As you go to ever increasing altitude, the weight of air above each successive level becomes smaller and, therefore, the pressure becomes successively lower.

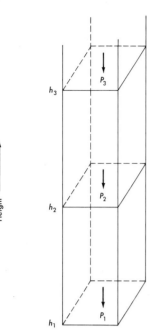

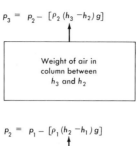

$$P_3 = P_2 - [\rho_2 (h_3 - h_2) g]$$

Weight of air in
column between
h_3 and h_2

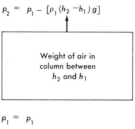

$$P_2 = P_1 - [\rho_1 (h_2 - h_1) g]$$

Weight of air in
column between
h_2 and h_1

$$P_1 = P_1$$

Figure 2-10 Pressure decreases with height and can be calculated from the hydrostatic equation knowing how density, ρ, varies with height, h. In this diagram ρ_1 and ρ_2 are the average densities in the height intervals $(h_2 - h_1)$ and $(h_3 - h_2)$, respectively. The acceleration of gravity, g, is constant.

Table 2-5 gives average values of temperature, pressure, and density for various altitudes between sea level and a height of 100 km at latitude 15°N. The distribution of temperature will be discussed in the next section. Both density and pressure decrease continuously as altitude increases.

Table 2-5 allows an answer to the question of how thick is the atmosphere. It is clear that there has to be a certain degree of arbitrariness in arriving at an answer because of the gradually decreasing density and pressure. It is seen that at 30 km the average pressure is about 12 mb and that 99 percent of the mass of the atmosphere is below that level. The 50-km level is above 99.92 percent and the 80-km level is above 99.999 percent of the atmosphere. Clearly, for most points of view the atmosphere can be regarded as being no more than 100 km thick, certainly a small value with respect to the radius of the earth, 6371 km. Meteorologists concerned with long-range weather prediction by means of mathematical treatments of the atmosphere concentrate on the region below 30 km.

TEMPERATURE STRUCTURE
OF THE ATMOSPHERE

As shown in Table 2-5, the average air temperature decreases with height in the lowest layers of the atmosphere. Before the days of high-altitude balloons it was thought that the temperature continued to decrease toward absolute zero in outer space. It came as a surprise at the turn of this century to find that at altitudes above about 10 km the air temperature remained nearly constant, or increased with height. It is now well known, as shown in Fig. 2-11, that the temperature distribution in the atmosphere consists of a series of relatively warm and cold regions.

As will be seen in the next chapter, most incoming solar rediation passes through the atmosphere and is absorbed by the surface of the earth. Heat from the ground is transferred to the air by conduction and radiation and is distributed through the atmosphere. Convection, which involves the ascent of warm air and the sinking of cold air, is particularly effective in transporting heat upward. As a result of these various processes, the average air temperature usually is highest near the ground and decreases with height until it reaches a level called the *tropopause*. It separates the lowest layer of the atmosphere, called the *troposphere*, from the next highest layer called the *stratosphere*.

In the interpretation of Fig. 2-11, one must keep in mind that these are average atmospheric conditions. In any particular place and time the temperature structure can be substantially different. Shallow layers are frequently observed in the troposphere where the temperature increases with height. They are known as *temperature inversions* because the change of

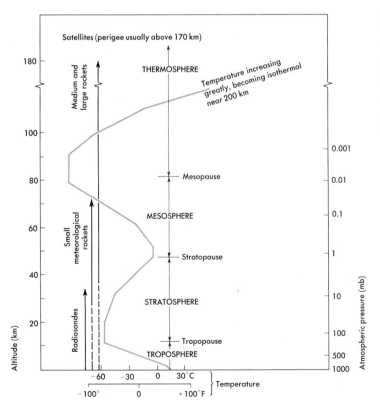

Figure 2-11 Average temperatures of the atmosphere as a function of height and the techniques used to measure them. The temperature structure defines the various regions identified on the drawing. *From* **R. S. Quiroz,** *Bulletin of American Meteorological Society,* 1972, **53**: 122–133.

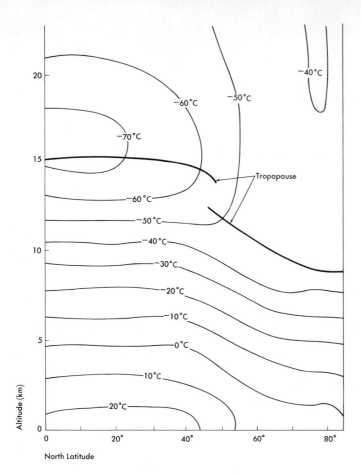

Altitude (km)

−40°C

−50°C

−60°C

−70°C

15

−60°C

−50°C

Tropopause

−40°C

10

−30°C

−20°C

−10°C

5

0°C

10°C

20°C

0

0   20°   40°   60°   80°

North Latitude

Figure 2-12 Average July tempera-
tures in the Northern Hemisphere.
Source of data: H. R. Byers, *General
Meteorology*, McGraw-Hill, 1974.

temperature with height is reversed from the more common decrease with
height.

In summer and at low latitudes the tropopause level is higher than
depicted in Fig. 2-11. This is illustrated in Fig. 2-12, which shows the height
of the tropopause as a function of latitude in the Northern Hemisphere in
July. On the average, the tropopause is at about 18 km over the equator and
8 km over the poles. At any one place it moves upward and downward with
the passage of warm and cold air masses.

In the lower part of the stratosphere the temperature is nearly
constant with height (i.e., it is *isothermal*) or it increases slowly. It increases
sharply in the upper part of the layer and reaches a maximum at the *strato-
pause* located at about 50 km. The warmth of the stratosphere results from
the absorption of ultraviolet radiation by ozone.* Although, as noted earlier,
the level of maximum ozone concentration is at 15 km to 30 km, small
quantities of ozone at altitudes of about 50 km absorb enough energy to
cause the observed high temperatures.

*Wavelengths between about 0.20 μm and 0.32 μm. The solar spectrum is discussed
in the next chapter.

When the atmosphere is isothermal or when the temperature increases with height, vertical motions of the air are suppressed. The atmosphere is said to be stable. In such a circumstance, atmospheric pollutants tend to persist for a long time. Since updrafts are inhibited, rain and snow clouds cannot form to wash out pollutants. When volcanic eruptions throw large quantities of particles into the stratosphere, the smallest ones may remain there for one to two years.

Supersonic transport airplanes, such as the Anglo-French Concorde and the Russian TU-144, are designed to fly at about 17 km and 20 km, respectively. That puts them into the stable stratosphere where the engine emissions can reside for long periods. As pointed out earlier, environmentalists are concerned that the emission of nitrogen oxides by large fleets of SST's will lead to a deterioration of the ozone layer and serious biological damage to people, animals, and plant life.

Through the *mesosphere* the temperature decreases with height up to about 80 km; this level is known as the *mesopause*. Above it there is a nearly isothermal layer some 10 km thick. Still higher, in the *thermosphere*, the temperature increases with height up to an altitude of about 200 km. The high temperatures are caused by the absorption of extremely shortwave ultraviolet radiation from the sun.* At the great altitudes of the mesosphere the air density is extremely low. Very little energy is needed to produce substantial temperature increases.

The absorption process in the mesosphere and thermosphere is called *photoionization*. Solar energies absorbed by molecules of nitrogen oxide (NO) and oxygen (O_2) or oxygen atoms (O) and other molecular species are sufficient to cause the ejection of an electron. Since remaining molecules and atoms are short one negatively charged electron, they are *positively charged ions*. Fig. 2-13 shows the distribution of ion concentration and the principal contributing atomic species in the atmospheric region called the *ionosphere*.

The ionosphere is divided into regions as indicated in Fig. 2-13. Since photoionization depends on the absorption of solar ultraviolet radiation, the ion concentration changes from day to night, particularly in the D and E regions. In the F region, where gas density is extremely low, the recombination of ions and electrons does not proceed as rapidly as it does at lower elevations. As a result, ion concentration in the F region does not change rapidly.

Before the days of artificial earth satellites the ionosphere played a vital role in long-distance radio communication. Depending on the electron concentration, it was generally possible to select a radio frequency that would be reflected by the E or F regions and allow radio contact over long distances, for example, New York to Paris. Under normal conditions, the

---

*Wavelengths of about 0.01 μm to 0.1 μm. The solar spectrum is discussed in the next chapter.

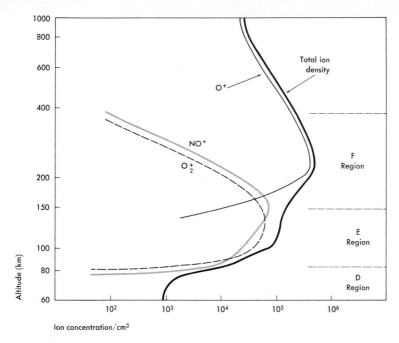

Figure 2-13 Average concentration of the principal, positively charged atomic and molecular ions at midday over middle latitudes. Note that the scales are logarithmic. *Source of data:* Air Cambridge Research Laboratories, Bedford, Mass.

concentration of electrons and ions in the D region is too low to interfere with the propagation of radio waves. Unfortunately, this is not always the case. Occasionally, the ion concentrations in the lowest two layers change rapidly and are accompanied by significant perturbations in long-wave radio signals.

Meteorites, most of them small particles of stone or metal from outer space, passing through the upper atmosphere at very high velocities cause sudden changes of ion concentrations in the E region. At the leading edge of meteorites, which may be only millimeters in diameter, there are high pressures and temperatures that cause the ionization of air molecules. The resulting increase of ion concentration can lead to sharp increases in the strength of reflected radio signals. Incidentally, meteorities range in size from the plentiful extremely small ones that pass through the atmosphere unaltered to the rare, huge meteorites that weigh several hundred kilograms. In total the fall of cosmic particles into the earth's atmosphere amounts to about 5 million metric tons per year.

The most serious interferences with radio propagation via the ionosphere are caused by solar disturbances. At periodic intervals there are huge eruptions from the surface of the sun. Most of the material returns to the body of the sun, but some ionized particles are ejected into space at very high speeds. Simultaneous with these corpuscular emissions from the sun are sudden increases of the short-wave ultraviolet radiation. The radiation traveling at the speed of light, 3×10^8 m/s (186,000 mi/sec) requires about

8 min to travel the roughly 1.5×10^{11} m from the sun to the earth and leads to marked increases in the ionization of the D region. As a result, radio waves can be absorbed before they are reflected; the consequence is a fade-out or interruption of long-distance radio traffic by way of the ionosphere.

Nowadays most intercontinental radio and television transmission is via satellite. Radio frequencies are selected that pass through the ionosphere undisturbed. By means of directional antennas, signals are beamed toward a satellite where they can be reflected toward a receiving antenna at a distant place. A more effective procedure is to equip the satellite with devices to receive the incoming signal, amplify it, and transmit a strong signal toward the receiving station.

The particles emitted by solar flares travel at speeds of about 1.5×10^6 m/s, much slower than the speed of the ultraviolet radiation, and arrive in the earth's atmosphere perhaps 30 hours or so after the flare on the sun. The particles are charged and are deflected toward the polar regions by the magnetic fields of the earth. This leads to *magnetic storms* which are major disturbances in the magnetic fields of the earth.

As the high-speed particles collide with air molecules, excitation occurs and electrons are freed. When de-excitation occurs and the air molecules return to their original states, light (mostly green and red in color) is emitted. Since this can take place over very large regions at the same time, the northern skies can become illuminated in spectacular displays of brilliant colors. This phenomenon is called the *aurora borealis* (Fig. 2-14) in the Northern Hemisphere and the *aurora australis* in the Southern Hemisphere.

Figure 2-14 Auroral band with ray structure. *Photo by* V. P. Hessler, University of Alaska.

Aurorae are produced at altitudes of about 100 km to 130 km. They take on many shapes but mostly are in the form of arcs, bands, rays, or massive curtains. On the basis of analyses of the color spectra of the aurorae, it has been concluded that oxygen atoms and molecules, and to a lesser extent, nitrogen, are mainly involved in the ionization.

Interestingly, there is a cyclical nature in the occurrence of solar flares and the resulting disturbances in the upper atmosphere. They fluctuate with a period of 11 years in association with a rise and fall in the number of sunspots on the surface of the sun. They are large, cool regions in the solar atmosphere which appear as dark spots and have been recorded for well over two centuries. Solar flares are relatively brief occurrences around the sunspot regions.

Over the years many people have sought to show that sunspot number not only correlates with ionospheric disturbances, but also with weather systems in the lower atmosphere. Certain investigators have found intriguing correlations between solar activity and various meteorological factors. Skeptics, however, have noted that no one has offered a convincing theory on how relatively small changes in incoming solar energy can affect air motions in the troposphere.

CHAPTER 3

Energetics of the Atmosphere

THE ATMOSPHERE can be regarded as a huge heat engine that is fueled by energy from the sun. Incoming radiant energy heats the atmosphere, which then converts the heat energy to kinetic energy and subsequently does work. In the physical sciences the term *work* has a specific meaning; it is done by a force acting over a distance. When a force, *F*, causes air to move a distance, *d*, we say that work has been done and its quantity is equal to $F \times d$. This concept will come up again later in this book.

When considering the earth and its atmosphere it is necessary to understand the properties of solar energy and how this energy enters the geophysical system. The planet earth is about 150,000,000 km from the sun and it orbits counterclockwise around the sun in a period of about 365 days. At the same time the earth rotates about its axis with a period of 24 hours.

A crucial point to recognize is that the axis of rotation of the earth is tilted in relation to the plane of revolution of the earth. This point is illustrated in Fig. 3-1. As the earth moves in its orbit around the sun, the earth's axis makes an angle of 66.5° with the plane of revolution. It is more common to say that the plane of the equator of the earth is inclined at an angle of 23.5° from the plane of the earth's orbit.

The earth's orbit around the sun is actually elliptical, but just barely. An examination of the axes of the ellipse reveals that the orbit is very close to a circle. The earth–sun distance is minimum on about January 3 and amounts to 147×10^6 km. This point on the orbit of the earth which is nearest the sun is called the *perihelion*. It is a term often used when describing the orbit of artificial satellites around the earth. On about July 3 the earth is farthest from the Sun, 152×10^6 km away, a point on the orbit called the *aphelion*. Because of the differences in distances, the earth intercepts about 6 percent more solar radiation in January than it does in July. Since in the Northern Hemisphere it is warmer in July than in January, it is evident that the variation of the distance between the sun and earth cannot account for the temperature differences between the summer and winter.

When considering incoming solar radiation it is convenient to think of the sun's position in relation to the earth's equator. As the year progresses, the sun moves southward from June 22 to December 21 and northward during the other 6 months. The sun reaches its most northward position on about June 22. This point in the orbit is known as the *summer solstice* in the

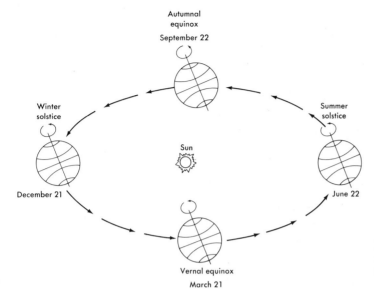

Figure 3-1 The seasons of the year can be explained in terms of the revolution, around the sun, of the earth rotating about an axis that has a fixed tilt in relation to the plane of the orbit.

Northern Hemisphere (Fig. 3-1). In the Southern Hemisphere June 22 is referred to as the *winter solstice*.

In the summer the rays of the sun strike most directly on the Northern Hemisphere, and as the earth rotates around its axis the heat is distributed around the earth. At the time of the summer solstice the Northern Hemisphere has its greatest exposure to the sun and the daylight is of maximum duration. At noon the sun is directly overhead at latitude 23.5°N. This latitude is called the *Tropic of Cancer*. North of the Arctic Circle, at 66.5°N latitude, the sun never sets and this fact is the inspiration for the phrase "the land of the midnight sun."

Six months later when the winter solstice occurs in the Northern Hemisphere, the sun's rays shine most intensely on the Southern Hemisphere. At noon on about December 22 the sun is almost directly overhead near 23.5°S latitude, the circle known as the *Tropic of Capricorn*. At this time of year the region north of about 66.5°N latitude, the Arctic Circle, is without sunlight. Except for twilight, the night period is 24 hours long. At the same time the area south of 66.5°S, the Antarctic Circle, is illuminated by sunlight for 24 hours.

As the sun moves northward after the winter solstice, the days become longer in the Northern Hemisphere. At the time of the *vernal equinox*, about March 21, the lengths of day and night are equal. On this date the sun at noon is just over the equator. The same is true some 6 months later on about September 22, the date of the *autumnal equinox*.

It is conventional to consider the seasons of the year as being decided by the solstices and equinoxes. For example, spring runs from about March 21 to June 22. These traditional designations have little meaning as far as weather-related activities are concerned. Farmers do not regulate their activities by this astronomical calendar. Nevertheless, the beginnings of the seasons do indicate changes in the lengths of day and night and, in a general way, represent gateways to changing climatic conditions.

In middle June the sun is almost directly above the Tropic of Cancer. Over the Northern Hemisphere incoming solar radiation intensity is higher than it is at any other time of the year. Why then are not air temperatures at a maximum during June instead of a month or so later, as is usually the case? The lag is explained by recognizing that the change of temperature of a volume of air depends on the difference between incoming and outgoing energy flux. As long as more energy comes in than goes out, the temperature continues to increase even if the incoming amount diminishes. This is the case in July in many regions and hence temperatures are warmer than they are in June. More will be said about this subject in Chapter 11.

THE NATURE OF RADIATION

An important property of radiation is its *wavelength*. When you see a wave motion on the surface of a body of water, perhaps produced by a stone thrown in a lake, it is easy to visualize the length of the waves. It is the distance from one wave crest to the next. Radiant energy is propagated through the atmosphere in the form of waves called *electromagnetic waves* because they can be explained in terms of electric and magnetic fields which exhibit alternate maxima and minima in field strength analogous to the crests and troughs of water waves.

Electromagnetic waves propagate through air at great speeds, at about 3×10^8 m/s. They have wavelengths ranging from extremely short X rays, through the visible wavelengths corresponding to the colors of the rainbow, the longer infrared band, and upward to radio waves (See Fig. 3-2).*

Before examining the nature of radiant energy from the sun and from the earth, it is helpful to review some important laws of physics. A point that is not commonly recognized is that all substances radiate energy unless they have temperatures at absolute zero. Since such a circumstance is never found on earth, we may state that everything radiates energy. The amount of this radiation can be calculated from a well-known law of radiation called the *Stefan–Boltzmann law*. It states that the maximum quantity of energy radiated in a unit time, from a unit area of a body having a Kelvin tem-

---

*When measuring very small distances, such as the wavelengths of light, it is common to use the units of micrometers (μm) or Ångstroms (Å). A micrometer equals 10^{-6} m and an Ångstrom equals 10^{-10} m. For example, the wavelength of red light is about 0.7 μm or 7000 Å.

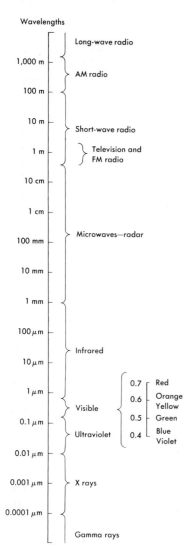

Figure 3-2 The electromagnetic spectrum showing the names given to different wavelength regions. Note the changes of units in the length scale.

perature, T, is proportional to the fourth power of T.* For example, if T increases from 300°K to 600°K, the radiated energy is increased by a factor of 16.

The actual amount of radiation depends not only on the temperature, but also on the nature of the substance. The maximum radiation at a given temperature is called the *blackbody radiation*. A rough, black surface emits more radiation than a highly polished, white surface. The term blackbody may

*The Stefan–Boltzmann law can be written as $E = \sigma T^4$, where E is the energy radiated in a unit time from a unit area, σ is a constant factor equal to 5.67×10^{-8} watts/m²(°K)⁴, and T is in degrees Kelvin.

be misleading, however, because a substance having this property might not be black or might not be a solid or liquid body. For example, snow acts nearly as a blackbody in the emission of infrared radiation. Water vapor is an efficient radiator at some wavelengths.

When considering the radiation from any substance it is necessary to recognize that it probably does not radiate exactly as a blackbody. This fact is taken into account by specifying a property called the *emissivity*. It is the ratio of the actual emitted power to that which would be emitted by a blackbody at the same temperature. The value of the emissivity, which ranges from zero to unity, depends on the properties of the substance and the wavelength of the radiation.

Another important property of any substance is its *absorptivity*, which is the ratio of the amount of radiant energy absorbed to the total amount incident on the substance. The absorptivity depends on wavelength and temperature. For any substance, the absorptivity is equal to the emissivity. This relationship is known as *Kirchoff's law*. Therefore, if an object effectively emits radiation of a particular wave-length, it effectively absorbs radiation at the same wavelength. The converse is also true. As a result, snow is a good emitter and absorber of infrared radiation, but it is a poor emitter and a poor absorber of visible radiation. For this reason snow is a good reflector of sunlight.

The wavelengths radiated by a blackbody depend on the temperature and can be calculated from *Planck's law*, one of the fundamental laws of physics. Starting from Planck's law it is possible to derive *Wien's law*, an expression that shows that the wavelength of maximum radiation intensity is inversely proportional to the absolute temperature.*

Consider some examples. The sun's radiation temperature is 6000°K. Wien's law shows that the wavelength of strongest intensity is 0.48 μm which is in the visible part of the electromagnetic spectrum. On the other hand, the average temperature of the earth is about 300°K and the wavelength of maximum radiation from the earth is 9.6 μm in the infrared region of the spectrum.

Most substances near the surface of the earth have temperatures between perhaps 220°K and 330°K, and therefore emit infrared radiation with energy maxima at wavelengths between about 9 μm and 13 μm. For example, consider people and the clothes they wear. The human body has a temperature of 98.6°F (37°C or 310°K). On the outisde of a person's clothing the temperature might be close to room temperature of 75°F (297°K).

According to Wien's law, a substance having a temperature between 297°K and 310°K radiates maximally a wavelength of about 9.5 μm. It is in the infrared part of the spectrum and is invisible to the naked eye. Therefore,

---

*$\lambda_{max} = 2880/T$. This is known as *Wien's law* and gives the wavelength of maximum radiation intensity as a function of absolute temperature. In this form, when T is in degrees Kelvin, λ_{max} is in micrometers.

if there were no source of visible light, you would not be able to see a person in the dark. Following Kirchoff's law, we can conclude that since humans effectively emit 9.5 μm radiation, they also effectively absorb radiation at this wavelength. On the other hand, since people do not emit visible radiation effectively, they also are not good absorbers of visible radiation. When light falls on a person, much of it is reflected and seen by an observer.

The effects of infrared radiation in transferring heat can be illustrated by considering what could happen to you in a house on a cold day. As already noted, the body temperature is about 310°K. When the outside air temperature is very low, the walls of a poorly insulated house might have a temperature of 62°F (17°C or 290°K). The radiation per unit area from the body would be greater than that from the cooler walls. The quantity of infrared radiation emitted by a person would exceed the amount of infrared energy absorbed, and therefore the body would lose heat to the walls. You could feel cool even though the *air temperature* in the room was at a supposedly comfortable level of perhaps 75°F. As this example illustrates, any real substance emits radiatiant energy and at the same time absorbs radiation. Its temperature changes as a result of radiation only if there is a difference between emission and absorption.

ENERGY
FROM THE SUN

Since the sun's radiating surface can be taken as having a temperature of 6000°K, the Stefan–Boltzmann law shows that the amount of energy radiated by the sun is enormous. Because the earth subtends such a small angle when viewed from the sun, the earth intercepts only about 0.5×10^{-9} of the sun's radiation.

The average rate of incoming radiation falling on the outer limit of the atmosphere at an angle perpendicular to a hypothetical sphere surrounding the earth is called the *solar constant*. Its value is taken to be about 2.0 langleys (abbreviated ly) per minute where 1 ly equals 1 cal/cm². One calorie is the quantity of heat required to raise the temperature of 1 g of water by 1°C. When the solar constant is multiplied by the cross-sectional area of the earth, it is found that the total solar energy intercepted is about 3.67×10^{21} cal/day. If this energy were *distributed uniformly over the surface of the entire globe*, the amount received per unit of area would be 263 kilolangleys per year (abbreviated kly/yr). Of course, the energy is not uniform over the earth; the areas near the equator get about 2.4 times more solar radiation than those near the poles.

Incidentally, other sources of energy for the atmosphere, such as heat from the interior of the earth, reflections of solar radiation off the moon, and energy from the solar tides, are much smaller than the sun's direct insulation. In total they are about 0.0002 of the solar constant, and therefore they can be disregarded when considering the energy budget of the global atmosphere.

Planck's law allows calculations of the spectrum of blackbody radiation from the sun. As shown in Fig. 3-3, the spectrum corresponds to a smooth, bell-shaped curve skewed toward longer wavelengths. Almost all of the energy is in the wavelength band between about 0.1 μm and 3 μm. This is often called *shortwave radiation*. A comparison of Figs. 3-2 and 3-3 shows that most solar radiation is at visible wavelengths. The existence of these waves is beautifully illustrated by the appearance of a rainbow in which raindrops serve as tiny prisms to separate the color components from red to violet.

As solar radiation crosses the enormous distance from the sun to the earth, it undergoes relatively little change in its spectral distribution except in the ultraviolet end of the spectrum. When the solar rays encounter the atmosphere, the gases and particles cause changes in the spectral properties of the incoming radiation as it propagates through the air.

As already noted in Chapter 2, electromagnetic radiation can be absorbed by various gases. Specific atoms and molecules absorb specific wavelengths. The ionosphere exists because of photoionization caused by extremely short, ultraviolet waves (0.01 μm to 0.1 μm long). The ozone layer is maintained by the absorption of ultraviolet radiation (wavelengths from about 0.2 μm to 0.32 μm). Other gases, particularly carbon dioxide and water vapor, are particularly effective absorbers at specific wavelength bands at wavelengths greater than about 0.7 μm. The absorptivities of the principal absorbing gases in the earth's atmosphere are shown in Fig. 3-4. As a result of absorp-

Figure 3-3 The spectrum of blackbody radiation from the sun calculated from Planck's law for a temperature of 6000°K.

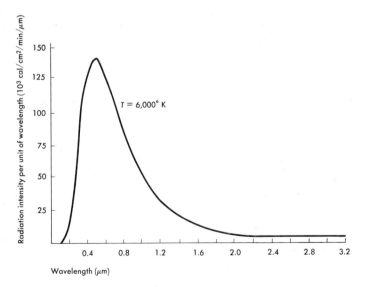

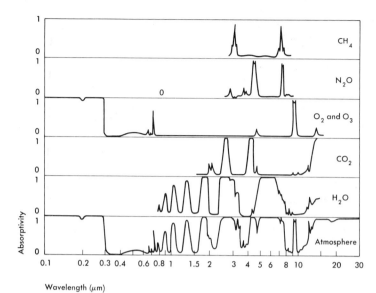

Figure 3-4 Absorptivities of the principal absorbing gases in the earth's atmosphere: methane, CH_4; nitrous oxide, N_2O; oxygen, O_2; ozone, O_3; carbon dioxide, CO_2; and water vapor, H_2O. The lowest diagram shows the absorptivity of moist air. *From* R. G. Fleagle and J. A. Businger, *An Introduction to Atmospheric Physics*, Academic Press, 1963.

tion corresponding to the bands in this illustration, the spectrum of solar radiation reaching the surface of the earth, known as *insolation*, is substantially different from the spectrum at the top of the atmosphere (see Fig. 3-5).

This diagram also shows the blackbody radiation calculated from Planck's law for a temperature of 300°K, a reasonable average for the earth as a whole. It is evident, as expected, that terrestrial radiation is in the infrared region. The strong and widespread absorption bands of water vapor and carbon dioxide prevent much of the terrestrial radiation from escaping to outer space. The absorbed energy, some of which is reemitted downward, serves to warm the lower atmosphere. These results explain why temperatures tend to be higher on humid nights than on dry nights. In desert areas the water vapor content of the air usually is low and as a result terrestrial radiation can escape readily to outer space. At night there is pronounced cooling. On the other hand, during humid, clear nights outgoing infrared radiation is inhibited from escaping because of water vapor absorption and temperature differences between day and night are small.

When clouds are present, the water droplets effectively absorb the terrestrial infrared and at the same time radiate infrared toward the ground. A cloud layer, especially one close to the ground, serves as a thermal insulator acting to reduce temperature changes of the air near the ground.

The curves in Fig. 3-5 show that most incoming solar radiation passes through the atmosphere and that a large fraction of the terrestrial radiation is absorbed by the air and radiated back toward the ground. This set of circumstances is called the *greenhouse effect*. In a glass greenhouse the short-

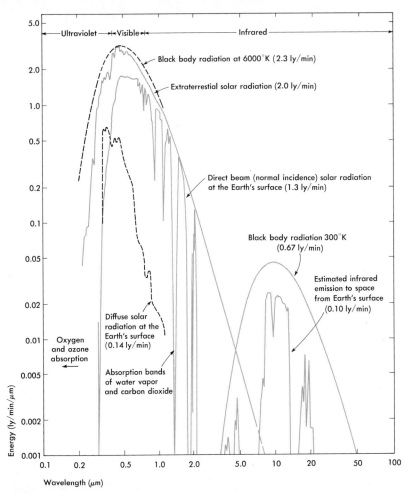

Figure 3-5 Calculated and observed spectra of radiant energy from the sun and the earth. The blackbody radiation, at 6000°K, is reduced by the square of the ratio of the sun's radius to the average distance between the sun and the earth in order to give the energy flux that would be incident on the top of the earth's atmosphere. *From* W. D. Sellers, *Physical Climatology*, University of Chicago Press, 1965.

waves from the sun pass readily through the glass while the infrared radiation emitted by the plants, surfaces, and air inside the house is partly absorbed by the glass. As a result, on sunny days air temperatures in greenhouses are very high. The same is true inside automobiles, particularly if they have large windows. It is important to realize, however, that a closed structure such as a greenhouse has a ceiling that prevents warm air from rising and leaving the greenhouse. In the atmosphere convection plays an important role in transporting heat upward away from the ground and thereby reducing temperatures.

As was mentioned in the last chapter, the combustion of fossil fuels has been causing a steady increase in the carbon dioxide concentration in the atmosphere. Figs. 3-4 and 3-5 show that carbon dioxide is transparent to solar radiation but not to terrestrial infrared radiation. The greenhouse effect of carbon dioxide acts to warm the lower atmosphere. Other factors cause opposite effects. More will be said about this matter in the discussion of global climate in Chapter 11.

In the absorption process, electromagnetic energy is intercepted by the gases and particles in the atmosphere and converted to heat. This process raises the temperature of the absorbing substance which also radiates infrared waves. Their wavelengths would depend on the temperature of the substance in accordance with Planck's law.

Air molecules and particles in the atmosphere also *scatter* electromagnetic waves. The electromagnetic energy incident on the molecules and particles is intercepted and redirected in many directions. A small fraction of the energy is backscattered toward the source of the radiation. This is sometimes called the *reflected* part of the scattered radiation. Radiation may also be scattered in the forward direction and all other directions as well.

The amount of scattering by a molecule or particle depends on its size, shape, and composition as well as on the wavelength of the radiation. When the scattering entities have diameters much smaller than the wavelength, the scattered power is proportional to the sixth power of the diameter and is inversely proportional to the fourth power of the wavelength. This is known as *Rayleigh scattering.** Consider how it applies to sunlight. The visible part of the solar spectrum runs from violet and blue where the wavelengths are about 0.4 μm to red where the wavelength is about 0.7 μm. Gas molecules have diameters of about 0.0001 μm and therefore they scatter radiation in accordance with Rayleigh's law. It shows that the smaller the wavelength the greater the scattering. Therefore, the same gas molecules will much more effectively scatter the violets and blues than they would scatter the oranges and reds.

As the sun's rays pass through the atmosphere, the blue and violet wavelengths are scattered in all directions while the reds and oranges pass through relatively unaffected. If you look, up, you can see bluish light being directed toward you from the countless number of scattering molecules. This fact explains the blue sky during days when the air is clean of polluting particles. Tiny particles of smoke, dust, and other substances also can be Rayleigh scatterers of visible radiation, providing that their diameters are much smaller than 0.4 μm.

Rayleigh scattering also explains the reds and oranges seen just at and after sunset and before sunrise. In these cases, the rays from the sun have had the blues and violets scattered out and only the longer waves have succeeded in passing through long atmospheric paths.

As a result of natural events such as volcanic eruptions, soil blown by

---

*The Rayleigh scattering law can be stated as $P = kD^6/\lambda^4$ where P is the scattered power, k is a constant, D is the particle diameter, and λ is the wavelength.

the winds and evaporating sea spray, as well as because of human activities involving combustion, locomotion, construction, and cultivation, the atmosphere sometimes becomes loaded with particles whose diameters can be as large as 5 μm to 10 μm. They clearly are too large to be Rayleigh scatterers and do not selectively scatter blue waves more effectively than the red ones. Instead, they scatter all the wavelenghts in the visible spectrum and the sky has a whitish appearance. As a rule of thumb, the bluer the sky, the less the quantity of atmospheric particles. In some large industrialized regions where sources of pollution are plentiful, deep blue skies are a rare occurrence.

The effects of volcanic eruptions can be interesting because they can inject huge masses of material into the atmosphere. A strong eruption can throw up as much as 100 billion cubic meters of fine particles. These substances and particles produced by condensed gases may ascend high into the stratosphere. The larger particles fall out fairly quickly even from the stratosphere, but the tiny ones can remain there for periods of a year or two, as is shown in Fig. 2-5. After the massive explosion of Krakatoa in Java in 1883 there was an increase of atmospheric turbidity much the same as the one following the Agung eruption noted in Fig. 2-5. The minute, persistent particles in the stratosphere have little effect on outgoing long-wave radiation from the earth, but they scatter incoming solar radiation and therefore cause small but important changes in the energy budget. The consequences on the global climate will be discussed in Chapter 11.

The extremely small, long-residing particles in the stratosphere can cause the selective scattering of various wavelengths in the optical spectrum. As a result, during sunset and sunrise on clear days the skies can take on beautiful colorations with deep hues of blue and lavender. They were especially brilliant in 1963 and 1964 after the eruption of Mount Agung, but they are seen after all major eruptions.

RADIATION BUDGET OF THE EARTH

The average atmospheric temperature of the earth varies slowly. Over a period of half a century the average air temperature near the surface may increase or decrease by 1°C. From certain points of view such a change is very important. It can have a profound effect on sea ice, sea level, and the growing season in marginal regions of the world. Nevertheless, when considering the overall heat balance of the earth, it is reasonable to begin with the assumption that the temperature is constant over a long period. In order for this to be true, the amount of incoming solar radiation must be balanced by the amount of outgoing radiation from the surface of the earth and the atmosphere.

The magnitude of the total incoming solar radiation and the total energies of various other phenomena and processes are given in Table 3-1. It is evident that most items on the list expend very small quantities of

Table 3-1 Approximate Total Energies of Geophysical Phenomena and Human Activities

Phenomena and activities	Calories
Solar energy received by the earth per day	3.7×10^{21}
World use of energy in 1950	10^{19}
Strong earthquake	10^{19}
Average cyclone	10^{18}
Average hurricane	10^{17}
Eruption of Krakatoa volcano, August 1883	10^{16}
Kinetic energy of the general circulation	10^{16}
Average squall line	10^{15}
Average magnetic storm	10^{14}
Average summer thunderstorm	10^{13}
Detonation of 20-kiloton Nagasaki bomb, August 1945	10^{13}
Average earthquake	10^{13}
Burning of 7000 tons of coal	10^{13}
Daily output of Hoover Dam	10^{13}
Average forest fire in the United States, 1952–1953	10^{12}
Average local shower	10^{11}
Average tornado	10^{10}
Street lighting on average night in New York City	10^{10}
Average lightning stroke	10^{8}
Individual gust of air near the earth's surface	10^{4}
Meteorite	10^{3}

Source: W. D. Sellers, *Physical Climatology*, University of Chicago Press, 1965.

energy in comparison to the total solar radiation incident on the top of the atmosphere. The table also shows that most human endeavors such as the generation of power at Hoover Dam and the explosion of nuclear weapons involve quantities of energy that are small in comparison with the energy from the sun. In later chapters we shall discuss most of the weather phenomena listed. The table gives a scale for judging the energetics of each one. For example, it can be seen that a lightning stroke represents only about 0.00001 of the energy in an average summer thunderstorm.

A substantial fraction of the solar energy incident on the top of the atmosphere is reflected back to outer space by air molecules and small dust particles, by clouds, and by the surface of the earth. The fraction, expressed in percent, is called the earth's *albedo*, and on the average is about 30 percent. Before weather satellites came into regular use, estimates of the albedo were based on inadequate global observations, and the widely used values were too high.

Consider what happens to the 263 kly/yr of solar radiation assumed to be uniformly distributed over the earth at the outer limits of the atmosphere. As illustrated in Fig. 3-6, 30 percent, or 79 kly/yr, are reflected back by clouds, the atmosphere, and the surface. Atmospheric gases, dust, and clouds absorb 50 kly/yr and the surface of the earth absorbs 134 kly/yr.

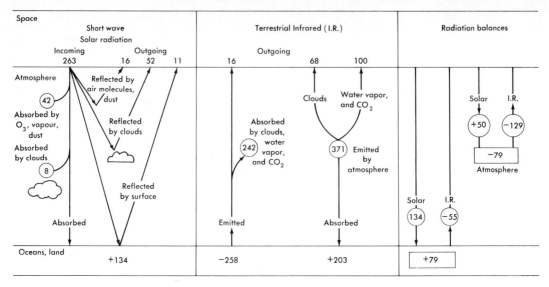

Figure 3-6 Average annual radiation budget of the earth. All units are in kilolangleys per year. Incoming solar radiation equals 263 kly/yr. *Based on data in Understanding Climate Change*, U.S. National Academy of Science, Washington, D.C., 1975.

In order to prevent a steady warming or cooling of the planet, the 184 kly/yr of energy absorbed by the atmosphere, the land, and the water of the earth must be returned to outer space. This occurs in the form of long-wave radiation emitted by the earth.

As shown in Fig. 3-6, the surface emits about 258 kly/yr of which 242 kly/yr are absorbed by atmospheric constituents such as water vapor, clouds, and carbon dioxide. These substances, in turn, emit about 371 kly/yr of which 203 kly/yr are absorbed at the surface of the earth. As a result, the net emission of infrared radiation from the earth's surface amounts to 55 kly/yr. Of the 371 kly/yr emitted by the atmosphere, a total of 168 kly/yr escapes to outer space (about 100 kly/yr from water vapor and carbon dioxide and 68 kly/yr^{-1} from the clouds). Thus, on the average, there is a balance because a total of 184 kly/yr of outgoing infrared radiation compensates for the same quantity of absorbed shortwave solar radiation.

On the average over the earth the surface absorbs 134 kly/yr of solar radiation while it radiates 55 kly/yr of infrared radiation. The difference, 79 kly/yr, is called the *radiation balance* or *net radiation* of the earth's surface. As noted in Fig. 3-6, the atmosphere absorbs only 50 kly/yr of solar radiation while radiating 129 kly/yr. Therefore, the net radiation of the atmosphere is –79 kly/yr.

These results indicate that energy is transferred mostly from the sun to the earth's surface and then to the atmosphere. In order to prevent the surface from getting progressively warmer and the atmosphere from getting progressively colder, there must be a steady transport of heat from the earth's surface (continents and oceans) to the atmosphere.

The radiation balances at the earth's surface are important quantities for anyone concerned with the development of techniques for the use of solar energy. For the earth as a whole, the average net radiation balance of 79 kly/yr corresponds to about 100 W/m². This means that an area 1 m² in size would be needed to keep a 100-W light burning even in the ideal case of a system with 100 percent efficiency in converting net radiation to electrical energy. If infrared emissions could be prevented and if all of the incoming solar radiation of 134 kly/yr could be collected, this would amount to about 170 W/m² on the average. At low latitudes in desert regions more appropriate values would be two or three times those given in this paragraph. These results indicate that the amount of available solar energy falling on a square meter of the earth is not particularly large, but if you used a large collecting area, the total available energy would be very substantial. For example, a collecting surface one kilometer square (10⁶ m²) could intercept about 500,000 kW.

Small solar collectors have been employed in many sunny places to heat water for domestic use. For heating buildings and for industrial use, it is necessary to develop economically feasible techniques for collecting a large quantity of energy, concentrating it, storing it, and utilizing it for space heating or power generation.

HEAT TRANSPORT
IN THE ATMOSPHERE AND OCEANS

In the preceding section we considered the energy balance averaged over the earth as a whole and found a positive net radiation at the surface that was balanced by a negative net radiation in the atmosphere. Such a balance does not exist at all latitudes. As shown in Fig. 3-7, the net loss of raidation by the atmosphere is almost the same at all latitudes, while the net gain of radiation by the surface is a maximum in tropical regions, decreases towards the poles, and becomes negative around the poles.

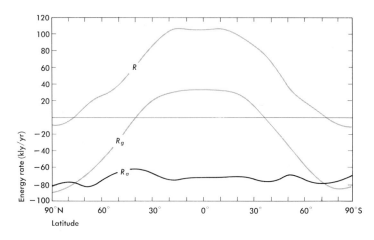

Figure 3-7 The average annual latitudinal distribution of radiation balances of the earth's surface R, of the atmosphere R_a, and the surface–atmosphere system R_g. *From* W. D. Sellers, *Physical Climatology*, University of Chicago Press, 1965.

Since incoming solar rays are more nearly perpendicular to equatorial areas than they are to polar areas, average insolation intensity is maximum at low latitudes and decreases with latitude. Note that the distance from the sun to the earth plays no part in explaining latitudinal variations of insolation.

Figure 3-8 depicts solar rays striking the earth at noon on about March 21. Because the sun is so far away, the rays can be regarded as being parallel. At the equator the rays are perpendicular to the earth's surface, that is, the angle between the rays and the surface is 90°, while at the poles the angle is 0°. At intermediate latitudes the rays intercept the surface at an angle which is 90° minus the latitude angle. In this drawing the rays are spaced uniformly so that we can imagine that there is an equal quantity of energy between any two rays. For example, the energy in the shaded regions labeled *E* and *P* are the same. Note, however, that the energy in *E* is distributed over a smaller area of the earth than is the energy in *P*. The dependence of incident radiation intensity on elevation angle is further illustrated in Fig. 3-9. The circular rays strike a surface at perpendicular incidence and at an angle of 45°. If the intercepted area at vertical incidence is *A*, the intercepted area is 1.4*A* when the ray strikes the ground at an angle of 45°. This means that if each ray contained the same quantity of energy, the energy per unit area at 45° incidence would be about 70 percent of that at vertical incidence.

In this discussion of solar radiation we have so far neglected to mention the effects of the atmospheric gases and particles. As noted earlier, they serve to absorb and scatter solar radiation, and as a consequence, the quantity reaching the earth's surface is less than that striking the top of the atmosphere. Figure 3-8 illustrates that near the equator the rays pass through a shorter atmospheric path than they do at more poleward latitudes. As a result, on the average, there are smaller losses of solar radiation in the equatorial atmosphere than in the polar atmosphere. This factor is a second reason why insolation at the earth's surface is highest over equatorial latitudes and decreases toward the poles.

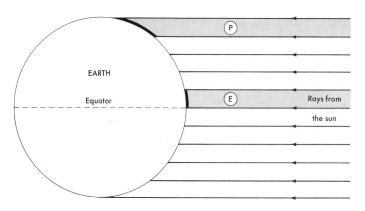

Figure 3-8   The energy per unit area is greater over the equator than at higher latitudes.

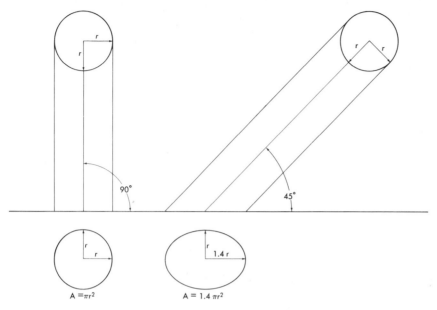

Figure 3-9 As the elevation angle decreases, the area over which the radiation is distributed increases.

The radiation balance of the earth–atmosphere system is positive equatorward of latitudes about 40° and negative at higher latitudes. In other words, the earth as a whole gains heat by radiation at latitudes equatorward of 40°, but it loses heat at more poleward latitudes. If there were not a transfer of heat poleward in both hemispheres, the tropics would get progressively warmer. Since that has not happened, there must be a transport of heat from low to high latitudes. The quantities of heat flow are shown in Fig. 3-10. In general, the maximum poleward heat flux occurs in the latitude band 40° to 50°.

Various forms of energy are transferred across latitude circles by air and ocean currents. *Sensible heat* is the energy represented by the motion of air molecules and is measured by the air temperature. As warm air and ocean currents flow poleward, while cold currents flow equatorward, there is a poleward transport of sensible heat.

Figure 3-10 shows that a substantial fraction of the energy transported in a north–south direction is in the form of *latent heat*. This term refers to the heat absorbed or released as a unit mass of water evaporates or condenses, respectively. The latent heat of vaporization depends on temperature and varies from 569 cal/g to 629 cal/g as the temperature goes from 50°C to −50°C. For most purposes, a value of 600 cal/g is a good approximation. When a gram of water evaporates, about 600 cal of heat are absorbed by the air and given up by the water surface. The loss of latent heat is readily experienced when you step out of a swimming pool on a warm, dry day. Even though air temperatures are high, the loss of heat as water evaporates from your body can cause you to feel chilled.

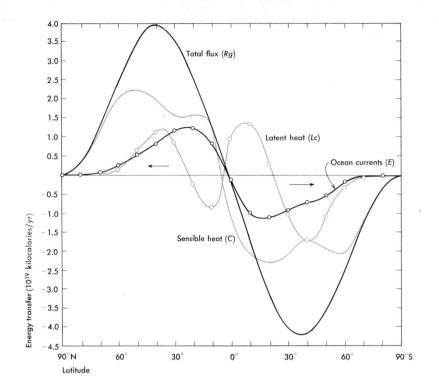

Figure 3-10 The average annual latitudinal distribution of the components of the poleward energy fluxes in units of 10^{19} kcal/yr. *From* W. D. Sellers, *Physical Climatology*, University of Chicago Press, 1965.

When water vapor condenses to form cloud droplets, latent heat is released and serves to warm up the air. If the droplets evaporate in the atmosphere, the atmosphere is cooled again. On the other hand, if the cloud droplets lead to rain and snow that fall to the surface, the atmosphere will have gained the released latent heat. Over the earth as a whole latent heat is transported from latitude belts where there is substantial evaporation to those latitude belts where there is substantial precipitation. The curves show that below about 22° latitude water vapor and latent heat are carried equatorward, while at higher latitudes they are carried poleward.

The major poleward energy flux is by means of air motions. At low latitudes, on the average, there is a convectively driven north–south circulation (Fig. 3-11). In such a circulation the air rises near the equator, moves poleward at high altitudes, sinks at latitudes of about 30°, and moves back toward the equator near the ground. Details of the air currents will be examined in Chapter 6.

In the middle latitudes the poleward transport of heat is by means of atmospheric disturbances that cause cold air from high latitudes to move equatorward while warm air moves poleward. At latitudes of 50° and 70° sensible heat flux accounts for most of the energy transfer.

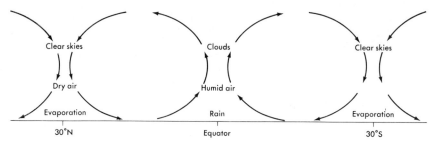

Figure 3-11 A schematic representation of the thermally driven longitudinal circulation at low latitudes. It is sometimes called the *Hadley cell*. See Chapter 6 for more details.

The oceans also play an important role in the poleward transport of heat. This is accomplished through the action of warm ocean currents traveling poleward while cold ocean currents move equatorward. When warm water moves under colder air it gives off energy in the form of latent heat and sensible heat which directly warms the air. Figure 3-10 shows how ocean transport of energy varies with latitude and how it is more important in the lower latitudes. Overall, ocean currents may account for 20 percent to 25 percent of the total poleward heat transfer.

AIR TEMPERATURES AT THE GROUND

Although heat is transported from equatorial to polar regions by various mechanisms, we know that, in general, temperatures are highest at low latitudes and decrease toward the poles. This point is illustrated in Fig. 3-12 that shows average air temperatures at *sea level* over the earth in January and June. If you were to draw a map of average air temperatures *at the ground*, there would be great variations caused by changes of station altitude, particularly in hilly and mountainous regions. Such a map would combine the effects of both horizontal and vertical displacements. In order to overcome this problem, climatologists construct maps of temperature at the same horizontal level, for example, at sea level. If an observing station is above sea level, its temperature is "reduced to sea level" by increasing it by an amount assumed to be reasonable considering average temperature changes with height. As a result, in Fig. 3-12 July temperatures over high regions such as central Mexico are indicated as being very high. In fact, because Mexico City is on a high plateau, actual surface temperatures in the summer are relatively cool. The chief purpose of the maps in Fig. 3-12 is to show how temperatures change over the earth as a whole. If someone wishes to know surface temperatures at any particular place not close to sea level, it is necessary to consult an appropriate climatological atlas (see Chapter 11 and Appendix III).

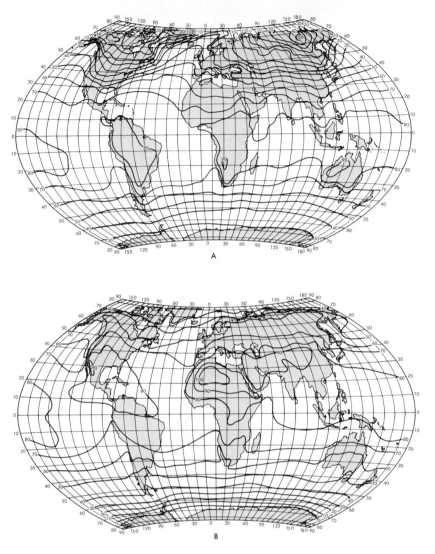

Figure 3-12 Average sealevel air temperatures in degrees Fahrenheit over the earth in January (A) and July (B). *From Climates of the World*, U.S. Government Printing Office, 1972.

Figure 3-12 illustrates a number of important facts. It shows that although air temperatures, on the average, decrease with latitude, the variations are not uniform. In the summer temperatures over the continents are higher than over the oceans. In the winter the reverse is the case. These results can be explained by a consideration of the properties of water and land.

In the first place, the albedos of the continents are different from those of the oceans. Earlier we defined the albedo of the earth as the fraction of incoming solar radiation that is reflected to outer space. In a similar fashion, you can define the albedo of any surface. Table 3-2 shows the albedos of various common surfaces found on the earth. It is evident that the fraction of incoming solar radiation reflected by the ground is dependent on the characteristics of the surface properties. Soil and vegetation absorb most of the incoming radiation. On the other hand, snow and ice regions reflect most incoming solar energy.

Table 3-2 ALBEDOS FOR SHORTWAVE
SOLAR RADIATION

Surface	Albedo (%)
Freshly fallen snow	75–90
Overcast of clouds	40–80
Sea ice	30–40
Dry sand	35–45
Wet sand	20–30
Dry clay or gray soil	20–35
Dark soil	5–15
Dry cement	17–27
Blacktop road	5–10
Deciduous forest	10–20
Coniferous forest	5–15
Crops	15–25

Source: W. D. Sellers, *Physical Climatology*, University of Chicago Press, 1965.

The amount of solar radiation absorbed by the oceans depends on the altitude of the sun. When it is less than 5° and the sun's rays are nearly parallel to a water surface, most of the insolation is reflected (see Fig. 3-13). As the sun's altitude increases, greater amounts of incident energy are absorbed. At the same time, the energy is distributed through a deeper layer of water. When the sun is directly overhead, direct solar radiation may penetrate to depths of more than 100 m. In clean sea water about 62 percent is absorbed in the first meter and 84 percent is absorbed in the first 10 m of water. The vertical mixing of water causes heat to be transported to much greater depths. Also, horizontal sea currents distribute heat through very large masses of water.

Solar radiation falling on land is absorbed in a thin layer of soil and rock—a few millimeters in fine-grained soil. Most of the energy goes to warm the top meter or so of the soil material. Because the solar energy on a square meter of ground is distributed over such a small mass of material, it produces a greater temperature increase than would the same amount of radiation energy falling on the oceans.

There is another important reason why land areas are warmer in the summer than are oceanic regions. It has to do with the *specific heat* of sand,

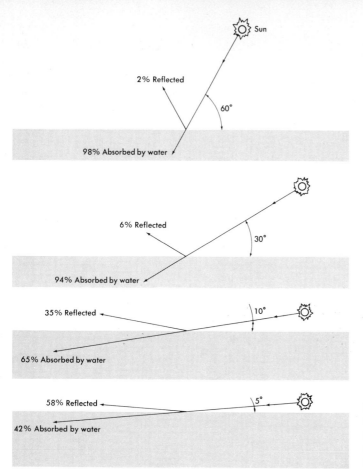

Figure 3-13 The amount of solar energy absorbed by a body of water depends on the angle of the sun's rays in relation to the water, that is, on the position of the sun.

soil, rocks, and water. The specific heat is the quantity of energy needed to raise a unit mass of a substance by 1°C. Table 3-3 shows that water has a specific heat of 1.0 cal/g°C, or, in other words, that the addition of one calorie of heat would raise the temperature of one gram of water by one

Table 3-3 SPECIFIC HEAT OF VARIOUS SUBSTANCES

Substance	Specific heat (cal/g·°C)
Air ($p = 1000$ mb; $T = 0°C$)	0.24
Quartz sand (medium fine, dry)	0.19
Ice ($T = 0°C$)	0.50
Granite	0.19
Sandy clay (15% water)	0.33
Calcareous earth (43% water)	0.53
Wet mud	0.60
Water	1.0

Source: R. J. List, *Smithsonian Meteorological Tables*, 6th ed., 1958.

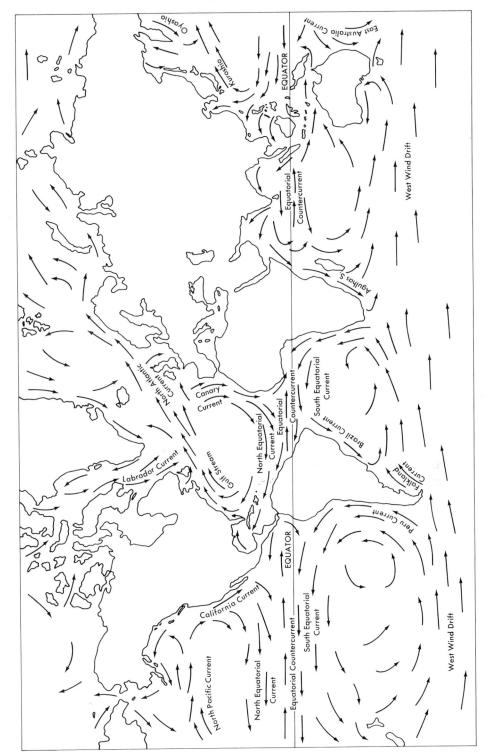

Figure 3-14 Major ocean currents of the earth. *Source:* U.S. Navy Hydrographic Office.

degree Celsius. On the other hand, 1 cal of heat would raise the temperature of 1 g of sand and granite by more than 5°C and the temperature of 1 g of clay by 3°C. Clearly, a given quantity of incoming solar energy would increase land temperatures more than it would increase the temperature of an equal mass of water. This also means that for a temperature reduction of 1°C, a mass of water gives off much more heat than does the same mass of sand or stone. It can be said that water has a relatively high heat capacity.

As already noted, heat from the sun is distributed through a large mass of ocean water, not only because of the penetration of solar radiation but also because of the overturning of water. The large mass coupled with the high heat capacity of water accounts for the fact that the oceans of the earth represent a huge reservoir of heat. When the air is cold, the oceans give off heat and act to warm up the atmosphere. When the air is warm, the oceans absorb heat. In this way, the oceans act as a huge thermostat as well as a heat reservoir. Sea temperatures change relatively little from summer to winter. This is in contrast to the continents whose temperatures change a great deal from the warm to the cold season.

Figure 3-12 also shows the effects of the major ocean currents on sea-level air temperatures. They can be understood by examining Fig. 3-14 and recognizing that, in general, poleward moving currents such as the Gulf Stream in the western North Atlantic and the Kuroshio Current in the western North Pacific are warm streams of water transporting heat toward the poles and warming the air. On the other hand, equatorward moving currents such as the California Current and the Peru Current off the west coast of North and South America, respectively, are cold streams of water. The air temperatures over the ocean currents tend to be high or low depending on whether the water is warm or cold respectively.

CHAPTER 4

The Winds

THE ATMOSPHERE IS A RESTLESS MEDIUM; it almost never stops moving. Sometimes air speeds are so low that flags hang limply from their staffs and anemometer cups on the roof of a weather station do not turn at all. But even on such occasions a sensitive measuring system would usually find some air movement. At the other end of the wind velocity spectrum one finds tornadoes and hurricanes in which the winds may exceed 100 m/s (224 mi/hr).

Although air motions are three-dimensional, it is convenient to examine the horizontal motions and the vertical motions separately. The upward and downward air currents are often called *updrafts* and *downdrafts*, particularly when considering conditions in clouds having strong vertical velocities. When meteorologists use the term "wind," they usually are referring to the horizontal components of velocity. The wind direction is always given as the direction *from which* it is coming. For example, when a wind vane shows the air moving from west to east, we say that a west wind is blowing.

There are many wind measuring instruments such as those shown in Fig. 4-1. Most people know that wind vanes are used to observe wind direction. Wind speeds often are measured by means of an anemometer having

Figure 4-1 (*Left*) A wind vane and three-cup anemometer and (*Right*) an aerovane. *Courtesy* Science Associates, Inc.

three or four hemispherical cups arranged around a vertical axis. The higher the wind speed, the faster the rate of rotation of the cups around the axis. Figure 4-1 also shows a device called an *aerovane* which measures both wind direction and speed. For many purposes, such as in the study of atmospheric turbulence, it is important to know three-dimensional air motions. In such cases, meteorologists often use a bivane, an instrument that points into the wind and can swing from side to side and up and down. By using a light, delicately balanced bivane, you can measure the rapid fluctuations of horizontal and vertical air motions. The value of such turbulence measurements will be noted in the next chapter.

FORCE, ACCELERATION, AND VELOCITY

The great English scholar Isaac Newton formulated a number of fundamental laws of physics dealing with the motion of bodies under the influence of various forces. According to Newton's second law of motion, if a force, F, is exerted on a body of mass, m, it undergoes an acceleration equal to the force divided by the mass.* Acceleration is the rate of change of velocity, that is, the change of velocity in a unit time. If, in a period of 1 sec, the velocity of an object changes from zero to 9.8 m/s, its acceleration is 9.8 m/s per sec which can be written 9.8 m/s^2.

At this point it is appropriate to remark that the quantities force, velocity, and acceleration are *vectors*. This means that in order to specify them completely, it is necessary to know both their magnitudes and their directions. For example, it is not sufficient to say that the wind velocity is 10 m/s. You also must give the direction from which the wind is blowing. Similarly, when you refer to a force, you must indicate the direction in which it is applied.

The concept of direction is especially important as it applies to acceleration. Newton's second law states that as the force changes, acceleration changes, and *vice versa*. It is easily seen that if a body of constant mass moves in a straight line, a change of acceleration must be accompanied by a change in the applied force. This accounts for the force on your body as an elevator or an automobile, moving in a straight line, speeds up or slows down. Note that the force is independent of velocity; as long as the motion is in a straight line, if the velocity is constant, there is no force regardless of the speed.

There is also an acceleration when the direction of motion changes even though the speed remains constant. Consider a car circling a track at constant speed. Since a velocity is specified by speed and direction, a person

---

*Newton's second law can be written $F = ma$, where F equals the net force, m is the mass of the system subjected to force F, and a is the acceleration.

in the car feels an outwardly directed force because the direction changes as the car moves in circular motion. The change in direction is an acceleration and is associated with a force.

Newton's second law applies to bodies of air as well as to other bodies. In order to explain the velocity of air, it is necessary to know the forces that cause air masses to accelerate.

There are a number of forces in action in the atmosphere. One of them, gravity, was mentioned earlier. The force of gravity is always directed downward toward the center of the earth and is equal to the mass of air times the acceleration of gravity (9.8 m/s²). In a unit horizontal cross section of the atmosphere the downward force per unit area is called the *atmospheric pressure.*

As a result of changes of temperature and to a lesser extent atmospheric humidity, barometric pressure varies from point to point both horizontally and vertically throughout the atmosphere. As was shown in Chapter 2, the hydrostatic equation allows calculations of the vertical distribution of pressure. As everyone who has looked at a weather map knows, there are regions of high and low pressure distributed over the face of the globe. The pattern of pressure is shown by the pattern of the *isobars*, lines along which the pressure is constant. Where the pressure differs from one point to the next on the same horizontal surface, there is a horizontal pressure force directed from high to low pressure. The magnitude of this force depends on a quantity known as the *pressure gradient*. It is the pressure change over a unit distance measured, in a horizontal plane, along the direction of maximum change. Along a line perpendicular to the isobars the pressure difference between two points divided by the distance between them is the pressure gradient (see Fig. 4-2).

If the earth were smooth, flat, and stationary, air would be accelerated from high to low pressure at a rate proportional to the *pressure gradient force*. Once the air begins to move, a second force comes into play—*friction*. It is a universal force found in every system involving movement. The rubbing of parts in any kind of machinery involves friction. Ball bearings and lubricating oils reduce frictional forces but they do not eliminate them. They act in a direction opposite to the direction of motion and they reduce speed. In the atmosphere there are frictional forces imposed on the air by the

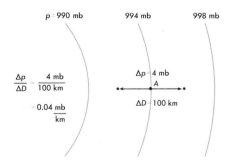

Figure 4-2 The pressure gradient is the change of pressure per unit of distance along a direction perpendicular to the isobars. In this example, the gradient at Point *A* is (996–992) mb divided by 100 km or 0.04 mb/km.

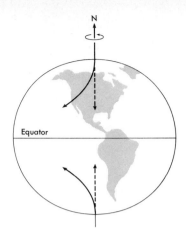

Figure 4-3 To an observer in outer space, missiles fired toward the equator from either pole would follow the straight dashed lines because of the absence of a force to change the trajectory. To observers on the moving earth, however, because the earth is rotating under them, the missiles will appear to follow the curved paths shown by the solid lines.

ground—rocks, vegetation, and other objects giving the surface a rough texture. In addition, a slow moving stream of air exerts frictional forces on faster moving air and acts to reduce its speed.

If pressure forces and friction were the only factors to be considered, it would be relatively easy to specify the movement of air in the atmosphere. Unfortunately, the specification of wind velocity is complicated by the fact that the earth is nearly spherical and rotates once every 24 hours.

In order to understand the effects of the earth's rotation, it is necessary to recognize that velocity is a *relative* quantity. For example, you might fly from Chicago to New York on a jet airplane that moves eastward over the ground at a speed of 300 m/s. At the same time the earth is rotating around its axis at a speed which measured at the equator amounts to 500 m/s while the earth as a whole is revolving around the sun at a very much greater speed. If you were on a distant star watching the jet airplane, its velocity would appear to be very different from what it appears when viewed from the earth.

The velocity of a body must be measured in relation to a specific reference point. The motion of air, that is, the wind, is measured in relation to the earth. Since the earth is moving at the same time that the air is moving, this convention can lead to difficulties.

Consider the case of a rocket fired southward from the North Pole (Fig. 4-3). According to Newton's first law of motion, once the rocket were fired, it would continue on a straight line unless another force is impressed to change its direction. From outer space the rocket would be seen to follow a straight course. As the rocket moves southward, the earth would be turning from west to east under the rocket. It would strike the ground at a point to the west of the point directly along the direction toward which it was launched. *To a person on the earth*, it would appear that the missile followed a trajectory curving toward the west. It would appear that there was a force acting on the missile and deflecting it toward the right. A rocket fired northward from the South Pole would curve toward the left when viewed from the earth.

The effects of the earth's rotation on the wind velocity is called the *Coriolis force* or the *Coriolis acceleration*, depending on the point of view of the writer. It is not a force in the sense of the gravity, pressure gradient, or friction forces. The concept of a Coriolis force is a complicated one that is not easy to explain or understand. The examples of the rockets fired from the Poles are straightforward, but it is more difficult to show that the Coriolis force affects the wind regardless of its direction and at all latitudes except the equator.

Figure 4-4 illustrates how the Coriolis force acts on a missile fired east northeastward from a point *P* taken to be at Chicago, Illinois. To an observer *in space* the missile follows a straight line (*P–F*). A person located at point *P* is carried around the earth and some time after the launch is located east of the original position. At Time 2 in Fig. 4-4 the line (*P–I*) shows the *initial direction* of missile velocity in relation to the earth, but the missile is moving in the direction (*P–F*). Because of the earth's rotation, the missile velocity appears to an observer at *P* to have been deflected to the right of its initial direction. The effect of the earth's rotation on the wind is the same as on the hypothetical missile. A similar analysis would show that in the Northern Hemisphere the Coriolis force causes a rightward deflection of wind velocity regardless of the direction.

If point *P* were on the equator, the direction of the missile flight would be the same whether an observer were on the earth or out in space. As the earth rotates, a north–south line through a point *P* on the equator would not rotate, as the 88°W longitude line does in Fig. 4-4. This means that the Coriolis force does not exist at the equator. The variation of Coriolis force with latitude can be explained by imagining that at every point on the spherical globe there is a vertical pole of rotation and that the local speed of rotation is some component of the speed of rotation around the North Pole. Since the North Pole is perpendicular to the equatorial plane, there is no component of rotation at the equator and therefore no Coriolis effect. It can be shown that the Coriolis force on a unit mass of air is equal to $V \times 2(\Omega$

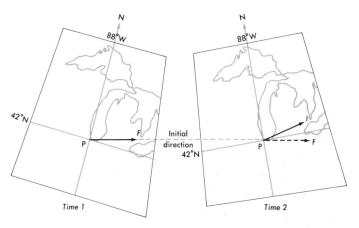

Figure 4-4 A missile fired eastward from point *P* appears to an observer at point *P* to be deflected to the right because of the Coriolis force. See the text for discussion.

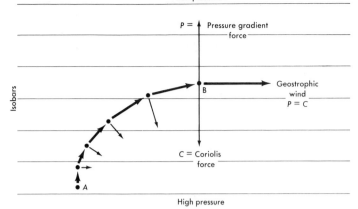

Low pressure

Figure 4-5 A schematic representation showing how a parcel of air initially at point *A* is acted on by the pressure gradient force (*P*) and the Coriolis force (*C*). As the wind velocity increases, the latter increases. At point *B*, *P* = *C* and the wind is parallel to the isobars and is called *geostrophic*. Note that there are no frictional forces.

sin ϕ) where V is the wind velocity, Ω is the angular speed of the earth (equal to $7.29 \times 10^{-5}/s$), and ϕ is the latitude. The quantity ($\Omega \sin \phi$) is the component of the angular speed at any latitude. The trigonometric function, sin ϕ, is zero at the equator and increases continuously with latitude being 1.0 at the poles. At latitudes of 30° and 60°, sin ϕ equals 0.50 and 0.87, respectively.

In summary, the Coriolis force varies in magnitude from zero at the equator to a maximum at the poles. In the Northern Hemisphere the Coriolis force acts to the right of the direction of the wind and in the Southern Hemisphere it acts to the left. As will be seen, this accounts for the fact that in the Northern Hemisphere winds blow counterclockwise around centers of low pressure but in the opposite direction in the Southern Hemisphere.

In order to illustrate the effects of the Coriolis force, let us consider a case in the Northern Hemisphere where the isobars are straight and parallel and the frictional forces are negligibly small. Imagine that the system starts from rest. At the outset there would be a pressure gradient force, *P*, causing the air to accelerate toward low pressure.*

As the air begins to move and the wind velocity, *V*, increases, the Coriolis force, *C*, increases and causes the wind vector to be deflected toward the right (Fig. 4-5). After a short period the deflection reaches the point where the wind velocity is parallel to the isobars. At this time there is a balance between the pressure gradient force, *P*, and the Coriolis force, *C*. When this condition applies, the wind is said to be *geostrophic*. Since there is no *net force*, there is no acceleration and the air moves at constant speed parallel to the isobars. As will be seen in the next section, such a state of affairs applies reasonably well in the upper atmosphere where frictional effects of the ground are vanishingly small.

*The magnitude of the pressure gradient force per unit mass can be written

$$P = \frac{1}{\rho} \frac{\Delta p}{\Delta D}$$

where ρ is the air density and Δp is the difference of pressure between two points separated by a distance ΔD (see **Fig. 4-2**).

The geostrophic wind velocity is proportional to the pressure gradient. This means that the closer the isobars, the greater the pressure gradient force and the higher the wind speeds. This analysis, as illustrated in Fig. 4-5, also shows that in the Northern Hemisphere low pressure is to the left of an observer looking downwind. This was recognized long ago and was given the name *Buys Ballot's* law. In the Southern Hemisphere low pressure is to the right of an observer looking downwind.

THE GRADIENT WIND

Even a casual examination of weather maps shows that there are regions of high and low pressure distributed over the earth. Often, as shown in Fig. 4-6, the low-presssure centers are nearly circular. As specified by Buys Ballot's law, in the Northern Hemisphere the winds around a low-pressure region tend to blow parallel to the isobars in a counterclockwise direction. The relationship of wind velocity and the pressure pattern in the Southern Hemisphere is illustrated by the weather map of Australia shown in Fig. 4-7. Note that, in general, the air moves clockwise around a low-pressure center and counterclockwise around a high-pressure center.

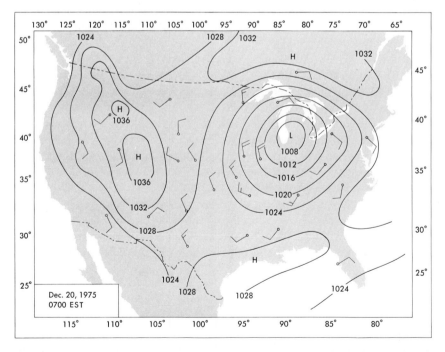

Figure 4-6   Sea-level isobars over the United States and observed wind velocities at selected stations.

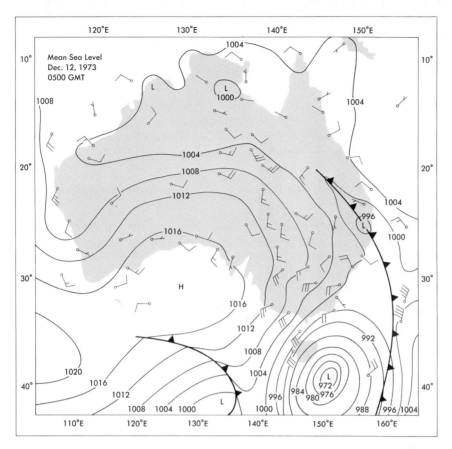

Figure 4-7 Sea-level isobars over Australia and observed wind velocities at selected stations. The heavy lines with arrowheads are fronts to be discussed in Chapter 7. *Courtesy Bureau of Meteorology, Melbourne, Australia.*

In order for the air to follow a circular path, there must be a net force directed toward the center. Figure 4-8 illustrates the forces acting on a small volume of air when *friction is so small that it can be disregarded.* As shown, the pressure gradient force is directed toward low pressure and the Coriolis force is to the right of the wind. In order for the wind to be parallel to the isobars and move in a circular path around the low pressure center, the pressure gradient force must exceed the Coriolis force.

If the pressure gradient force and the Coriolis force were equal in magnitude, there would be no net force, and the air would move in a straight line in accordance with Newton's first law of motion. The reason the air follows the circular path is that the pressure force toward low pressure exceeds the Coriolis force. It can be said that this net force causes a centripetal acceleration, that is, an acceleration inward toward the center of rotation. When friction is negligibly small the difference between the pressure gradient force and the Coriolis force accounts for the centripetal acceleration, and the wind, called the *gradient wind*, is parallel to the isobars.

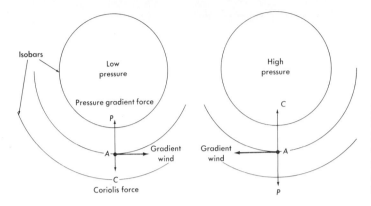

Figure 4-8 Wind velocities around low- and high-pressure centers in the Northern Hemisphere in the absence of friction. The force directed toward the center of the circular isobars exceeds the outward directed force and the air follows the circular isobars.

Around a high-pressure region in the Northern Hemisphere the balance of forces leads to gradient winds that blow clockwise, that is, *anticyclonic*. As depicted in Fig. 4-8, for the same pressure gradient, the wind speeds are higher around a high pressure center. This accounts for the fact that the Coriolis force exceeds the pressure gradient force and there is a net force toward the center of curvature of the isobars.

EFFECTS OF FRICTION

Close to the ground the effects of friction can cause the winds to deviate markedly from the ideal patterns shown in Fig. 4-8. This is particularly true in mountainous areas where the terrain can have profound influences on wind velocities up to great altitudes.

As noted earlier, friction is a universal force that is always present to a certain degree whenever there is motion. It acts in a sense opposite to the direction of motion and therefore is a decelerating force. Near the ground frictional effects on the wind are pronounced. By reducing the wind speed, friction also leads to a reduction of the Coriolis force, but it does not affect the pressure gradient force. As shown in Fig. 4-9, the net force resulting from the vector addition of the three acting forces leads to a wind velocity that is not parallel to the isobars. Instead, the wind is deviated across the isobars toward the lower-pressure regions. Winds have components toward a center of low pressure and out of a center of high pressure. These points are illustrated by the observed winds shown on the sea-level weather maps in Figs. 4-6 and 4-7.

At upper levels of the atmosphere the frictional effects of the ground are small, and the winds are much more nearly parallel to the isobars. This is clearly seen in Fig. 4-10, which displays wind velocities at an altitude of about 5.5 km at the same time as the surface map shown in Fig. 4-6. The solid lines shown in the diagram are equivalent to isobars in pattern but they are not isobars. In order to explain this point, it is necessary to digress briefly from the discussion of winds.

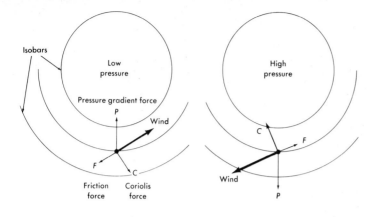

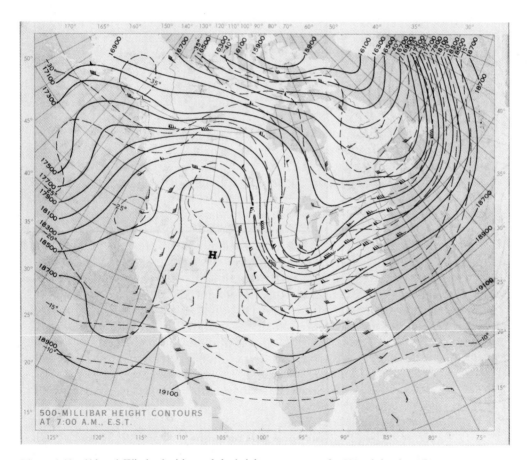

Figure 4-9 (Left) Wind velocities around low- and high-pressure centers near the ground in the Northern Hemisphere when frictional forces, *F*, are important. The effect of friction is to reduce wind speed and cause a deviation of the wind toward lower pressure. As a result, air blows into low-pressure centers and out of high-pressure centers.

500-MILLIBAR HEIGHT CONTOURS AT 7:00 A.M., E.S.T.

Figure 4-10 (Above) Wind velocities and the height contours on the 500-mb level at 0700 EST on December 20, 1975 (the same time as the surface map shown in Fig. 4-6). The heights are in feet; to convert to meters, multiply by 0.3048. The dashed lines are temperatures in degrees Celsius. Each wind barb corresponds to 10 knots; each arrow head to 50 knots. One knot equals 0.51 m/s.

At one time meteorologists studied weather patterns in three dimensions by drawing isobars at various levels in the atmosphere: sea level, 3 km, 5 km, etc. About three decades ago it was found more convenient, when mathematically describing the atmosphere, to accomplish this objective by depicting the heights of surfaces of constant pressure instead of the pressure at a constant height surface. Now this practice is used all over the world. If you have never encountered this notion, it may be difficult to visualize. Remember that when the pressure in one region is high in relation to the surrounding areas, the height of a constant pressure surface is also high in relation to the surrounding area. Consider for the moment the level in the atmosphere that corresponds to a pressure of 500 mb. You might recognize that since the average sea-level pressure of the atmosphere is 1013 mb, the 500-mb surface divides the total mass of the atmosphere roughly in half. On the *average*, the 500-mb surface is at an altitude of about 5.5 km above sea level, but it varies from less than 5 km to more than 6 km.

The map in Fig. 4-10 shows that on that day the height of the 500-mb surface varied from as low as 4860 m in the region of lowest pressure to as high as 5821 m in the region of highest pressure. When you examine this and other 500-mb charts, you should interpret the lines of constant height as if they depicted pressure variations. The closer the lines of constant height, the closer would be the isobars, the greater would be the pressure gradients, and the stronger would be the wind velocities. Note that through the central United States, where the lines of constant height are close together, the wind speeds are high. Maximum speeds are shown to be about 95 knots* (49 m/s). Observe how nearly parallel the wind vectors are to the lines of constant height. This illustrates that frictional forces are very small. At these levels the wind is well represented by the gradient wind and can be calculated accurately if the pressure pattern is known.

As already noted, winds in the lowest layers of the atmosphere are markedly influenced by friction. The ground, vegetation, buildings, etc., act to slow the motion of the air moving over them. This frictional effect is distributed upward as a result of the viscosity of the air. In essence, this means that the molecules of slow-moving air in one layer mix with the molecules in an adjacent layer of fast-moving air. As a result the initially low speeds are increased and the initially high speeds are decreased.

Another means by which frictional effects are propagated upward from the ground is by the vertical movement of *eddies*. Eddies are small volumes of air—picture them as bubbles if you like—that move from one altitude to another. Ascending eddies of air transport slow-moving air upward and descending eddies transport fast-moving air downward. More will be said about eddies in the next chapter.

---

*In the United States on weather maps above the surface, wind speeds are plotted in knots (1 knot equals 1 nautical mile per hour).

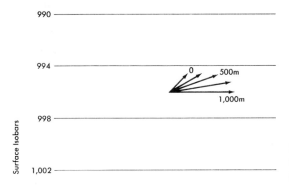

990

994

998

1,002

Surface Isobars

0   500m

1,000m

Figure 4-11 The Ekman spiral is the name given to the turning of the wind through the friction layer from the ground to a height of about 1000 m.

As already stated, close to the ground the winds blow slightly across the isobars toward low pressure. At the gradient wind level, at perhaps 1000 m, where frictional forces are negligibly small, the winds are nearly parallel to the isobars. This means that through this frictional boundary layer the wind direction turns with height. At the same time the wind speed increases with height because of the smaller retarding effect of friction. Figure 4-11 illustrates how the wind velocity vector turns to the right with height until it becomes gradient, that is, parallel to the isobars. The variations of wind with height through the friction layer is sometimes called the *Ekman spiral* after the man who first developed a theory to explain its existence.

The boundary layer of the earth's atmosphere is of crucial importance because we live in it and breathe it. Much more will be said about it in later chapters.

LOCAL WINDS OF SPECIAL INTEREST

In many places around the world there are wind currents of a local nature that are difficult to explain on the basis of the pressure patterns observed on conventional weather maps. The winds occur, at least in part, because of topographical peculiarities. Winds that have unusual properties and occur often enough have been given names by the people who experienced them. Winds having the same physical properties may have different names depending on the country involved. A few examples of commonly observed local winds are worth noting.

Along coast lines it is common to experience *sea breezes* on summer afternoons (Fig. 4-12). A sea breeze occurs because solar radiation increases the temperature of the land and the air just above it more than the temperature of the air over the water. As a result, a low-level pressure gradient develops from water to land. Cool air from over the sea (or lake) moves over

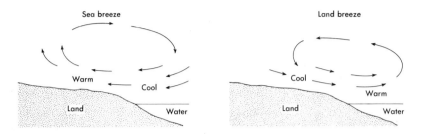

Figure 4-12   Sea breeze and land breeze circulations during the day and nighttime respectively.

the land and bathes the coast with gentle breezes. Over the land the air rises, moves out to sea aloft, and sinks toward the water in a form of a convection cell. At night the land cools faster than the water and the sea-breeze circulation is sometimes reversed. When this occurs, a light wind from land to sea develops and is called a *land breeze*.

In mountainous areas, particularly in regions where clear skies prevail, it is common to find mountain and valley winds. During the day the sun's rays heat the air over the mountain slopes. As the air warms, it becomes buoyant and flows upslope, that is, up the valley. At night heat is radiated to outer space, the air cools, becomes more dense, and a mountain wind flows downhill. When skies are clear and the air is dry, heating and cooling by radiation can be very effective in causing mountain and valley breezes. For example, in Tucson, Arizona, where these conditions are common, the afternoon wind blows upslope, that is, from the northwest on more than 50 percent of the days. During the cool, early morning hours the wind blows from the southeast (downslope) on about 80 percent of the days.

Some unusual circulations of a local character result from the interaction of topographical effects with certain large-scale features of the atmosphere. The warm, dry winds, called *foehns* in Europe, occur in such a fashion. They are frequently observed in the Alps, but they are also common on the east slopes of the Rocky Mountains, particularly in Wyoming and Montana where they usually are called *chinooks*. They may occur over any mountain range if the appropriate conditions occur.

Foehn winds are initiated by the advance of an upper-level, low-pressure center. When it passes over a mountain ridge, a low-level pressure trough is formed on the lee side of the mountain, and strong downslope winds develop on the lee side. The sinking air is warmed by compression in a manner described in the next chapter. The warming leads to a reduction in the relative humidity. In winter a foehn wind can bring welcomed warming, but sometimes the hot, dry, gusty winds reportedly cause psychological reaction such as irritableness and nervousness.

Since the foehn air replaces much colder air, there can be a sudden, intense warming. It has been reported that on one occasion, at Havre, Montana, a chinook caused the temperature to increase from 11°F to 42°F (−11.7°C to 5.6°C) in three minutes. Increases of air temperature of 10°C to 20°C in 15 min are not unusual.

The high temperatures coupled with low humidities can lead to rapid melting and evaporation of snow. It has been reported that 30 cm of snow can disappear in a few hours when strong chinooks occur. This explains why the Blackfoot Indians call them "snow eaters."

During some times of year the hot desiccating winds increase the danger of fire in forests. The Santa Ana of southern California is a foehn-like desert wind that blows, sometimes very strongly, from the deserts to the east of the Sierra Nevada Mountains.

Several other warm winds have become reasonably well known. The *sirocco* is observed in advance of low-pressure systems moving across North Africa. It is a warm, dry, dusty wind moving from the south or southeast across the Sahara. As the air moves over the Mediterranean Sea, it picks up a great deal of moisture because of its high temperature. When the sirocco reaches Malta and southern Italy it is hot, humid, and uncomfortable. A wind similar in nature to the sirocco blows northward over Egypt and is called the *khamsin*. Another variety of strong, very dusty wind observed in desert areas is called the *haboob*. The winds, which are created by air flowing out of the bases of large thunderstorm systems, can attain velocities of about 30 m/s. The turbulence and high velocities pick up large quantities of fine soil particles and carry them up to perhaps a kilometer. An advancing haboob has the appearance of a whirling wall of sand. These winds are well known in the Sudan, but they also occur in the Sonoran desert of southern Arizona.

There are a number of well-known cold winds. A *blizzard* is not really a wind; it is a weather condition with strong winds, low temperature, and enough snow to restrict visibility to below about 150 m. According to the U.S. National Weather Service definition, a *severe blizzard* is present when the wind speeds exceed 20 m/s, temperatures are below about −12°C (10°F), and there is enough snow in the air to reduce visibility to nearly zero. In Russia and Siberia a cold, snow-filled northeasterly wind in winter is called a *purga*.

The *bora* is another cold, gusty wind that occurs in various parts of the world. The name was originally applied to northeast winds on the Dalmatian coast of Yugoslavia. The current originates as very cold air from Russia, crosses the mountains, and descends to the relatively warm shores of the Adriatic Sea at speeds sometimes exceeding 40 m/s. Although the air in a bora is warmed as it sinks, it starts off with such low temperatures that it still is cold when it reaches low elevations.

The *monsoon* circulation over India and southeast Asia is crucial to the survival of millions of people (Fig. 4-13). During the winter when a high pressure prevails over Asia air streams southward over India and neighboring countries. This is known as the winter monsoon and is characterized by relatively cool, dry air.

As the summer approaches, the Asian continent begins to warm and low pressure develops. This leads to the development of a broad stream of humid air from the Indian Ocean flowing northeastward. The air rises as it passes over the rising terrain approaching the Himalaya Mountains. In normal circumstances this leads to heavy showers and thunderstorms and torrential rain. For example, Cherrapungi, India, has an average annual rainfall of *11.4 m*, mostly between June and the end of September. When the summer monsoon is delayed or is much weaker than normal, the rice fields are

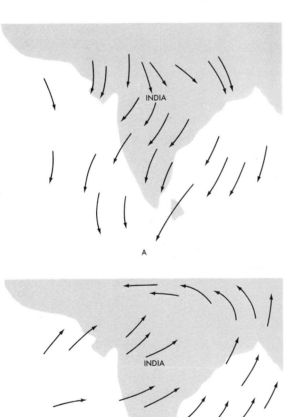

Figure 4-13   Wind patterns over India during the winter monsoons in January (A) and summer monsoons in July (B).

inadequately watered, agricultural production suffers, and great hardships are visited on many people.

In Chapter 6 we shall deal with the general circulation of the atmosphere. It will be seen that the change of wind pattern with season, which is the chief characteristic of a monsoon, also occurs over the United States and elsewhere.

CHAPTER 5

Atmospheric Stability and Vertical Motions

VERTICAL MOTIONS IN THE ATMOSPHERE can be produced in a number of ways. Some are obvious. For example, air ascends when it is forced to move over rising terrain. When the direction of flow is over a gradually sloping region such as the Western Great Plains, the upward velocities are fairly small, perhaps 10 cm/s. On the other hand, if air blows against a steep mountain barrier, its vertical velocities can amount to several meters per second.

Air is also lifted over weather fronts. As will be seen in the next chapter, when large bodies of cold and warm air come into contact, they do not mix readily. Instead, the cold, heavier air moves in wedge-like fashion under the warmer, less dense air. The transition between the warm and cold air is called a *front*. As the cold air advances, the warm air it displaces is caused to rise.

In the preceding chapter it was shown that because of frictional effects, winds near the ground are deviated toward low pressure. As a result, the air converges toward low-pressure centers, and it rises because the ground blocks downward flow (Fig. 5-1). If the air is assumed to be incompressible, which can be taken to be the case in treating most problems involving air motions, low-level convergences cause updrafts. Diverging air in the lower layers of anticyclones is accompanied by sinking air.

Meteorologists sometimes refer to the vertical velocity of the air produced by the terrain or by some dynamical process as being *mechanically* induced. A crucial consideration is whether the air accelerates or decelerates after it has started to move vertically. As was the case with horizontal motion, the answers are supplied by Newton's second law of motion that equates net force to acceleration. Since atmospheric pressure decreases with height, there is an upward-directed pressure-gradient force while the force of gravity is directed downward. The difference between these two forces is referred to as *buoyancy* or a *buoyancy force*. In the simplest terms, when a volume of air is warmer than its surroundings, the upward pressure-gradient force exceeds the downward gravity force, there is positive buoyancy, and the air rises; when a volume of air is colder than its surroundings, it is negatively buoyant and it sinks. To be more exact, when considering buoyancy, it is necessary to consider the air density instead of the temperature. Water vapor reduces the density of air because the molecular weight of water vapor is 18 while

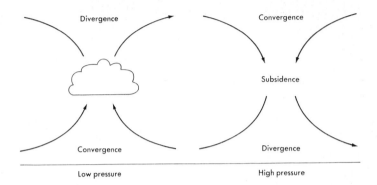

Figure 5-1 Patterns of vertical air motions, convergence and divergence, associated with low- and high-pressure centers.

air behaves as if it were a gas having a molecular weight of 28.9. As a result, at any temperature the more humid the air, the less its density and the greater the buoyancy force. Data in Table 5.1 show the effects of humidity on air density when the air temperature is 10°C and pressure is 700 mb. In general, humidity differences are less important than temperature differences in determining the buoyancy of a volume of air.

When considering the buoyancy of air containing cloud droplets, ice crystals, and raindrops, it is necessary, for precise work, to determine how their masses affect the total mass of a unit volume of cloud. Liquid or ice particles increase the overall density of the volume of air in which they exist because the densities of water and ice are about 1000 kg/m³ while the density of air ranges from about 1.2 kg/m³ at sea level to 0.3 kg/m³ at 12 km. The total masses of liquid or ice particles are small in relation to the mass of air in the same volume. For example, at an altitude of about 2 km a cubic meter

Table 5.1   DENSITY AND VIRTUAL TEMPERATURE OF AIR
AT VARIOUS RELATIVE HUMIDITIES*

Relative humidity (%)	Air density (kg/m³)	Virtual temperature (°K)†
0	0.861	283.0
25	0.860	283.5
50	0.858	283.9
75	0.857	284.4
100	0.856	284.9

*Density and virtual temperature calculated at an atmospheric pressure (p) of 700 mb and a temperature (T) of 283°K (10°C).

†The virtual temperature is calculated from the equation

$$T_v = \frac{T}{1 - \frac{3}{8}\frac{e}{p}}$$

where T is the actual temperature in degrees Kelvin, e is the vapor pressure.

of air has a mass of about 1000 g while the mass of water in liquid and frozen form in a typical cloud is almost always less than 10 g/m³. Nevertheless, the relatively small masses of water may have significant effects on the buoyancy forces, particularly at higher altitudes.

STABILITY AND INSTABILITY

Imagine a small volume of air beginning to rise as it passes over a hill. Meteorologists use the term *air parcel* to designate such a small imaginary body of air having uniform physical properties. If, after ascending to a certain level, such as the altitude 2 shown in Fig. 5.2, the air is less dense than the air in the surroundings, it will be subjected to an upward buoyancy force.

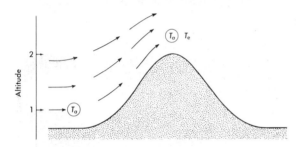

Figure 5-2 Air moving over a hill or mountain ridge is forced to rise. If the temperature, T_a, of the rising parcel is higher than the temperature, T_e, of the environment, the parcel is accelerated upward.

As a result, there will be an upward acceleration of the already rising air. By means of an application of Newton's second law, it can be shown that the upward acceleration is proportional to the temperature difference between the parcel and the environment, that is $(T_a - T_e)$ in Fig. 5-2.*

As already noted, the buoyancy force also depends on the water vapor concentration in the air, but usually to a lesser degree than it depends on temperature. The effects of humidity can be taken into account by employing a quantity called the *virtual temperature*. It is the temperature of dry air having the same density and pressure as a system of moist air. The virtual temperature is higher than the actual temperature of the moist air because both increased humidity and increased temperature lead to decreased density. Table 5-1 gives an example of how the virtual temperature of air,

*The vertical acceleration a is given by

$$a = g \times \frac{T_a - T_e}{T_a}$$

where g is the acceleration of gravity and equals 9.8 m/s², T_a is the temperature of the rising or descending air, and T_e is the temperature of the environmental air expressed in degrees Kelvin.

having an actual temperature of 283 °K (10°C), is increased as the relative humidity is increased.

Since temperature differences usually are much more important in determining buoyancy than are moisture differences, we shall disregard the latter in most discussions of vertical motions in the atmosphere. This is done to reduce the complexity of the subject.

When a rising parcel of air is warmer than the surrounding air, it is accelerated upward and the atmosphere is said to be *unstable*. In a more general sense, an unstable state exists when a volume of air, displaced from one level to another and then released, continues to accelerate in the *direction of the displacement*. When the air is unstable, vertical air velocities both up and down are accelerated by buoyancy forces. Stability with respect to vertical motions is more precisely referred to as *hydrostatic stability*.

Under certain atmospheric conditions, when a volume of air is lifted to a higher altitude, it arrives there colder and more dense than the surrounding air. In this case, there is a downward buoyancy force and the volume of displaced air is forced back toward the level of origin. In such circumstances, the atmosphere is *stable*.

If a lifted volume of air arrives at a new altitude having the same density as the surrounding air, there will be no net buoyancy force. The volume will remain at the new level, and the atmosphere is said to have *neutral stability*.

TEMPERATURE LAPSE RATES
AND STABILITY

In the last section the meaning of atmospheric stability was discussed in terms of vertical accelerations and velocities. It is reasonable to ask what atmospheric properties determine stability. In this section it will be shown that stability depends mostly on the vertical distribution of air temperature.

The rate of *decrease* of temperature with height is called the *temperature lapse rate*. When temperature decreases with height, the lapse rate is positive. As shown in Fig. 2-11, this is the case in the troposphere. On the average, the temperatures at sea level and at an altitude of 10 km are 15°C and −50°C, respectively. Therefore, the average lapse rate is 65°C divided by 10 km or 6.5°C/km (3.6°F/1000 ft).

Before going on, it should be noted that in more specific terms the quantity 6.5°C/km is the *average environmental lapse rate;* it was obtained by using values of temperature averaged over time and space. The decrease of temperature at any given time and place is also called the environmental lapse rate. It needs to be distinguished from the *process lapse rate* that refers to the change of temperature of a given parcel of air as it rises or sinks in some atmospheric environment. It will be seen later that the chief criterion

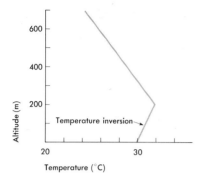

Figure 5-3 At night when the sky is clear and air is dry, a temperature inversion frequently forms near the ground.

for atmospheric stability is the difference between the environmental and the process lapse rates.

The environmental lapse rate can be very different from 6.5°C/km at any time or place. When the temperature increases with height, the layer through which this occurs is known as a *temperature inversion* and the lapse rate is negative. In the example in Fig. 5-3 the lapse rate through the inversion layer is −2°C/200 m or −10°C/km. If the temperature is constant with height, the lapse rate equals zero and the layer is described as *isothermal*.

The reason why stability depends on the temperature lapse rate can be seen by examining what happens when a parcel of air rises or sinks in the atmosphere. As it ascends from altitude 1 to altitude 2 in Fig. 5-4, it moves to a region of lower pressure. In the process, the volume of the parcel expands until the pressure inside equals the pressure outside it. A balloon rising in the atmosphere similarly expands as pressure forces act outward on the skin of the balloon. According to the laws of thermodynamics, the expansion requires that *work* be done because air is being moved by a force. The energy to do the work is supplied by energy extracted from the air molecules in the parcel. As a result, the temperature of the rising parcel decreases even though no heat is carried away from the expanding air. This is called *adiabatic cooling*, and the rate of decrease with height of the temperature of the ascending air is called the *adiabatic lapse rate*. It is a process lapse rate because it specifies the temperature change of a given parcel of air as it changes altitude.

The rate of cooling in an adiabatic process depends on the specific heat of the gas in the atmosphere and the acceleration of gravity. It is different on Venus, for example, than on Earth.

In the earth's atmosphere the adiabatic lapse rate of dry air is 10°C for each kilometer of ascent. A descending volume of dry air is warmed at a rate of 10°C/km. As air moves to lower altitudes and higher pressure, it is compressed, work is done on the air, and it is converted to sensible heat that causes an increase in the air temperature. If you know how the temperature of a rising parcel of air changes, it is possible to establish criteria for deciding if the atmosphere is stable or unstable.

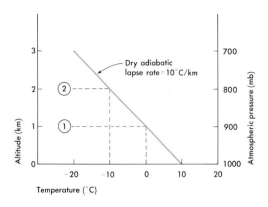

Figure 5-4 Rising dry air cools at the adiabatic rate of 10°C/km of altitude change. Sinking dry air warms at the same rate.

Figure 5-5 gives an example of a temperature structure on a day when the air is unstable. The lapse rate is 20°C/km. Consider the air originally at altitude A. If for some reason it were displaced to altitude B, it would follow the dashed line representing the dry adiabatic lapse rate. In other words, it would cool off at the rate of 10°C/km. When it reached B, the parcel would find itself warmer and less dense than the environment and would continue rising. If the parcel were moved to altitude C, it would be cooler and more dense than the environment and would continue to descend.

Figure 5-6 shows what happens in the case of an inversion. Again the rising parcel of air moves dry adiabatically. In this case, when it reaches altitude B, it finds itself colder and more dense than the environment. It sinks back toward altitude A along the dry adiabat. If the parcel were displaced to altitude C, it would become warmer and less dense than its surroundings. It would become buoyant and would return to altitude A. This illustration shows why temperature inversions are stable and suppress mixing.

Figure 5-5 When the environmental lapse rate is greater than the adiabatic lapse rate, the atmosphere is unstable.

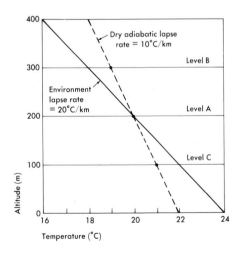

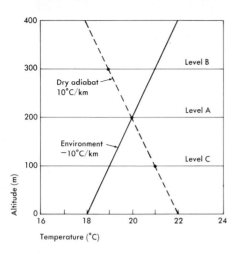

Figure 5-6 When the environmental lapse rate is smaller than the adiabatic lapse rate, the atmosphere is stable. This is very much the case when a temperature inversion exists.

As already mentioned, lapse rates in the atmosphere extend over a large range of values. For this reason, stabilities also vary greatly. Diagrams of temperature versus height, such as those shown in Figs. 5-5 and 5-6, can be used to determine whether or not the atmosphere is stable or unstable and to determine the degree of either condition. The boundary line between stability and instability is the dry adiabatic lapse rate. When the environment lapse rate is greater than adiabatic, the atmosphere is unstable. When the environment lapse rate is smaller than adiabatic, the atmosphere is stable. When the environment lapse rate is exactly equal to the adiabatic lapse rate, the atmosphere has *neutral* stability. If air is displaced vertically, it arrives at a new altitude with the same temperature and density as its surroundings. As a result, there is no net vertical force on the parcel and it remains at the new altitude.

TEMPERATURE INVERSIONS

Since temperature inversions represent a high degree of atmospheric stability, it is important to know where and how they form. Inversions near the ground often occur as a result of the cooling of the earth's surface by radiation. When the atmosphere is clear and dry, the earth begins to cool in the late afternoon when the outgoing radiation exceeds the incoming radiation. After sunset, radiative heat loss proceeds rapidly. This leads to a cooling of a shallow layer of air near the surface and the formation of a temperature inversion.

In a desert locality such as Las Vegas, Nevada (average annual rainfall of about 9.6 cm), ground-level temperature inversions are common during the night and early morning hours and are rare during the warm part of the day. As can be seen in Table 5-2, in the winter, when the sun rises late and sets early, temperature inversions persist in the mornings and start to form early in the evenings.

Table 5-2   PERCENT FREQUENCY OF TEMPERATURE
INVERSIONS BELOW 150 m
AT LAS VEGAS, NEVADA

	Time (PST)			
	0400	*0700*	*1600*	*1900*
Winter	92	92	2	88
Summer	89	41	1	54

Source: C. R. Hosler, Low-Level Inversion Frequency
in the Contiguous United States. *Monthly Weather Review*,
1961. **89**: 319–339.

In some cases, low-level temperature inversions continue for days. Usually, this occurs when warm air passes over a cold surface. For example, inversions form just above the ground when tropical air from the Gulf of Mexico moves northward over the United States during the winter. Inversions also develop as warm air moves over colder water. When this happens, the air loses heat to the underlying surface and results in a persistent, low-level temperature inversion.

Various types of temperature inversions are consistently observed in the atmosphere. At high levels there is the stratosphere (Fig. 2-11) which, because of its marked stability, places a lid on the growth of giant thunderstorms. Powerful updrafts sometimes penetrate the tropopause with vertical velocities that might exceed 30 m/s, but they are rapidly decelerated by negative buoyancy forces encountered in the stable stratospheric air.

Since thunderstorms and other precipitating systems do not extend significantly into the stratosphere and the stability suppresses mixing between troposphere and stratosphere, pollutants introduced into the stratosphere by volcanic eruptions, nuclear explosions, or aircraft remain for a long time. This point is illustrated by the data in Table 5-3 giving estimated residence times of particulate pollutants in various layers of the atmosphere. Following

Table 5-3   RESIDENCE TIMES OF PARTICULATE
POLLUTANTS IN THE ATMOSPHERE

Layer	*Residence times*
Lower troposphere	1–3 weeks
Upper troposphere	2–4 weeks
Lower stratosphere	6–12 months
Upper stratosphere	3–5 years

*Source: Man's Impact on the Global Environ-
ment*, Report at the Study of Critical Environmental
Problems (SCEP). M.I.T. Press, 1970.

the eruptions of powerful volcanoes such as Mount Agung in 1963, massive quantities of dust particles were injected into the stratosphere. They increased the overall turbidity of the atmosphere, as shown in Fig. 2-5, and escaped from the atmosphere very slowly.

Temperature inversions are common throughout the troposphere. As will be seen in Chapter 7, a frontal zone separating the cold air under a front from the warmer air above it generally is an isothermal or an inversion layer. The stability of the zone acts to suppress the mixing of cold and warm air and serves to maintain the front.

Many inversions in the lower atmosphere are attributable to sinking air. As noted earlier, when air subsides, its temperature increases at the adiabatic rate, 10°C/km. In some instances, the subsidence continues to a particular level and then the air diverges horizontally. A temperature inversion is commonly observed in the layer separating the upper region of sinking air and the lower region where there is almost no net vertical motion. Subsidence inversions are frequently found in regions of high pressure because they are characterized by sinking air. During the night in such a system there may be radiation inversion at the ground as well as a subsidence inversion aloft (Fig. 5-7).

As will be seen in the next chapter, there are large, persistent high-pressure regions, called *anticyclones*, centered over latitudes of about 30°N and 30°S (Fig. 6-2). In the eastern portion of these high-pressure regions,

Figure 5-7 Occasionally, a subsidence inversion occurs aloft and a radiation inversion occurs near the ground.

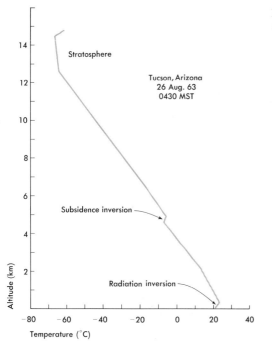

that is, near the west coasts of continents at latitudes near 30°, the subsidence is particularly pronounced. It leads to strong temperature inversions that usually are at about 500 m to 1000 m above sea level. The subsiding air and the inversions inhibit the formation of clouds and rain. This accounts for the extensive desert areas found under the eastern portions of the low-latitude anticyclones.

The high-pressure region dominating the eastern North Pacific Ocean extends over the southwestern United States and explains the prevailing sunny skies. It also accounts for a persistent subsidence inversion at an altitude of about 700 m. Under the inversion, pollutants from motor vehicles and industry become concentrated and lead to the famous Los Angeles smog.

TURBULENT DIFFUSION

In the discussion of stability we referred to rising and sinking parcels of air. Such parcels of variable size, each one having uniform properties, are sometimes called *eddies*. They carry along the properties of the air they had at their original locations. For example, an ascending eddy is likely to be transporting more polluted air than is a descending one (Fig. 5-8). As a result, pollutants are carried from regions of high concentrations to regions of low concentrations. This transport process is called *eddy diffusion* or *turbulent diffusion*.

The rate of diffusion can be related to the character of the eddies. They represent turbulent motions in the air that can be measured by means of a special wind vane known as a *bivane*. It responds very rapidly to changes in the horizontal wind as well as the vertical air velocities. As eddies of varying sizes and strengths are carried past the instrument by the average wind, the bivane is swung right and left, up and down. The more turbulent the air, the more frequent and extreme the swings of the vane.

The rates of turbulent diffusion of any atmospheric property, such as a polluting gas or aerosol, can be related to the fluctuations of the wind and the strength of the average wind. A plume of smoke from a chimney will be caused to spread vertically and horizontally depending on the vertical and horizontal wind fluctuations.

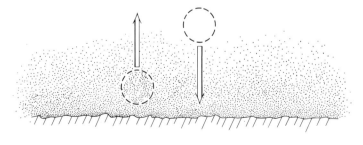

Figure 5-8 Upward and downward moving eddies act to transport and diffuse pollutants through the atmosphere.

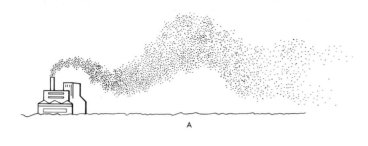

A

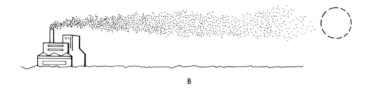

B

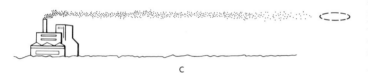

C

Figure 5-9 The character of a smoke plume depends on the turbulence properties of the atmosphere and the average wind speed. The more unstable the atmosphere, the greater the diffusion.

The degree of turbulence in the atmosphere depends primarily on the stability and the properties of the wind. As was shown earlier, the more rapid the decrease of temperature with height, the more unstable the air and the greater the vertical motions. During the daytime, when incoming solar radiation is greatest, the air tends to be unstable, and the lighter the wind, the greater the turbulence. Eddy sizes increase with altitude and so does the rate of spreading of a plume of contaminants. At night low-level temperature inversions tend to prevail. If they are accompanied by light winds, diffusion rates will be particularly small and pollutants emitted from ground sources will accumulate in the inversion layer. Therefore, pollutant emissions should be kept to a minimum during the night.

Most often eddy diffusion serves as a process for carrying pollutants upward from the ground, but this is not always the case. Sometimes, when smokestacks are the sources, diffusion transports the effluents downward. Because smokestacks are used often for the disposal of noxious gases and aerosols, there has been a great deal of study of how smokestacks function under various meteorological conditions.

The behavior of a smoke plume coming out of a high chimney depends mostly on the strength of the wind and the stability of the atmosphere. Three common classes of plumes are illustrated in Fig. 5-9.

When the wind is light to moderate but the air is unstable, a *looping plume* occurs [Fig. 5-9(A)]. It is characterized by a distinct wavy appearance that is produced by large eddies in the form of convective cells. Sometimes parts of the plume brush the ground close to the stack. Another common feature of a looping plume is that turbulent diffusion causes it to spread out rapidly, both horizontally and vertically. This result follows from the fact that as the air becomes more unstable, there is an increase in turbulence.

When the winds are greater than about 10 m/s, and the atmosphere is of neutral stability, *coning* takes place [Fig. 5-9(B)]. This means that the smoke spreads out in a form resembling a cone with only small upward and downward movement of the plume axis.

On occasions with low winds and a stable atmosphere, *fanning* is observed [Fig. 5-9(C)]. This means that the smoke remains in a fairly shallow layer, but it spreads out laterally as it moves downwind. It develops the appearance of a fan. Conditions for fanning occur most often during the night and in the early morning when the lower atmosphere is very stable and vertical motions are suppressed.

It is not unusual for the diffusion conditions to change from one type to another. For example, a fanning plume of the type just mentioned can be rapidly converted to a looping plume. Early in the morning the smoke may be trapped in a shallow, stable layer perhaps a hundred meters thick. After the sun rises and begins to warm the ground, the temperature structure and stability of the atmosphere begin to change. Close to the ground air is heated and becomes unstable and then it rises and mixes through a deeper and deeper layer. When the depth of the mixing layer reaches the level of the smoke plume, there is a rapid downward spread of the smoke. This result is sometimes called *fumigation*.

When a smokestack is being designed, it is important to know how high to make the stack in order to minimize the quantities of smoke reaching the ground. As you would expect, the higher the stack, the smaller the concentration of smoke at the ground. In general, as the stack height increases, the maximum ground-level concentration decreases as the square of the height. Thus, if the height is double, the concentration is reduced by ($\frac{1}{2} \times \frac{1}{2}$), or one-quarter. It has been found that the higher the stack, the greater is the distance downwind affected by smoke pollutants.

Of course, there are exceptions to these general rules. In cases of extreme looping, smoke from even the highest stack might reach the ground in high concentration. Furthermore, this might occur fairly close to the stack. When the atmosphere is stable, however, little smoke will be carried down to the ground even from short stacks. In mountainous terrain there are special problems because the pollutants from even the highest stacks, those about 300 m tall and located in valleys, can affect people, animals, and vegetation on the slopes at higher elevations.

This discussion sketches some of the factors governing the effectiveness of smokestacks. The purpose has not been to cover the subject in any detail, but rather to emphasize the importance of taking atmospheric conditions into account. It is equally important to consider them in determining the most favorable periods for releasing smoke into the atmosphere.

VERTICAL MOTION
OF MOIST AIR

Up to this point we have dealt with the motion of cloud-free air. When either condensation or evaporation accompanies vertical air motions, it affects buoyancy and vertical acceleration. This is the case because condensation and evaporation involve the exchange of latent heat, which, in turn, affects the temperature of the rising or sinking air. Before examining the effects of phase changes of water from vapor to liquid to ice, and *vice versa*, we must consider how humidity is described and represented.

The most common term used to designate the moisture content of the air is the *relative humidity*. The relative humidity can be defined in a number of ways, all of which consist of a ratio expressed in percent. The ratio represents the quantity of *water vapor* in a volume of air to the quantity that would be present if the air were saturated. Consider for a moment what is meant by the term *saturation*. It can be illustrated by imagining the sequence of events in a simple experiment. On a day when the air is dry, a small amount of water is poured into a jar and the jar is tightly sealed. At the outset the water vapor content of the air in the jar is small, but it increases with time as the water evaporates. At the surface of the water, water molecules escape into the air while a smaller number return to the water. As time goes on, as shown in Fig. 5-10, the quantity of water vapor in the air space within the jar approaches a leveling-off point. Finally, a point is reached when the number of evaporating molecules exactly equals the number of condensing molecules. The quantity of water vapor in the air remains constant and saturation has been reached.

If the substance in the jar were ice instead of water, the same sequence of events would occur. Water molecules escaping from the ice would exceed the number depositing on the ice until the air became *saturated with respect to ice*. At that point the number of molecules of H_2O leaving the ice surface and entering the surface would be equal.

The quantity of water vapor in the air can be measured in a variety of ways. *Absolute humidity* is the water vapor density; it is the mass of water vapor in a unit volume. The absolute humidity at saturation depends only on temperature, and as shown in Table 5-4 and Fig. 5-11, it increases with

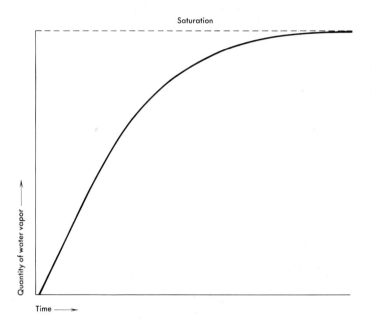

Figure 5-10 In a closed container holding water, evaporation continues until the air is saturated, at which point the number of water molecules leaving the water surface equals the number entering it.

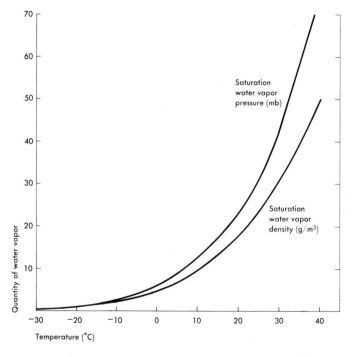

Figure 5-11 Variations with temperature of saturation values of water vapor pressure and vapor density. For example at 20°C the saturation vapor pressure is about 23 mb and the saturation vapor density is about 17 g/m³.

increasing temperature. These displays also show that the *saturation vapor pressure* does the same. Vapor pressure is the pressure exerted by the water vapor molecules and is expressed in millibars. If there were a flat plate in the gas, the water vapor molecules would bombard the plate and exert a force on it. The force per unit of area would be the vapor pressure.

The quantity of water vapor in the air can also be represented by the *mixing ratio*. It is the mass of water vapor in a unit mass of dry air. As shown in Table 5-5, the mixing ratio at saturation depends on both pressure and temperature. At any temperature it increases as pressure decreases, and as would be expected from the data in Table 5-4, the mixing ratio increases markedly with increasing temperature.

Table 5-4 SATURATION WATER DENSITY—SATURATION ABSOLUTE HUMIDITY AND SATURATION VAPOR PRESSURE

Temperature* (°C)	Vapor density (g/m³)		Vapor pressure (mb)	
	Over water	Over ice	Over water	Over ice
−40	0.176	0.119	0.189	0.128
−30	0.453	0.338	0.509	0.380
−20	1.074	0.884	1.254	1.032
−10	2.358	2.139	2.863	2.597
0	4.847	4.847	6.107	6.107
+10	9.399	—	12.27	—
+20	17.30	—	23.37	—
+30	30.38	—	42.43	—
+40	51.19	—	73.78	—

*Values at intermediate temperatures can be obtained by constructing smooth curves.
Source: R. J. List, *Smithsonian Meteorological Tables*, 6th ed., 1958.

Table 5-5 SATURATION MIXING RATIO OVER WATER
(Grams of water per kilograms of dry air)

Pressure (mb)	Temperature (°C)				
	−40	−20	0	20	40
1000	0.118	0.785	3.839	14.95	49.81
800	0.148	0.980	4.802	18.79	63.49
600	0.197	1.306	6.415	25.29	N.L.
400	0.295	1.960	9.664	38.70	N.L.
200	0.589	3.929	19.62	N.L.*	N.L.
100	1.179	7.903	40.49	N.L.	N.L.

*N.L. indicates values not listed in reference.
Source: R. J. List, *Smithsonian Meteorological Tables*, 6th ed., 1958.

According to international agreement, the preferred definition of relative humidity is the ratio, expressed in percent, of the actual mixing ratio to the saturation mixing ratio. It is also given, to a close approximation, by the ratio, in percent, of the actual to the saturation amounts of absolute humidity or vapor pressure. Because of its ready application to the physics of cloud droplet formation, the relative humidity is often written as the ratio of the actual vapor pressure to the saturation vapor pressure.

There are a variety of instruments, called *hygrometers*, that can be used for measuring atmospheric humidity. Devices for measuring and recording humidity are named *hygrographs*. One of the oldest measurement techniques uses clean, human hairs. As the relative humidity increases, the hairs get longer by a small but detectable amount. Blond hairs, 15 cm long at a relative humidity of zero, can increase to about 15.4 cm at 100 percent. In a hair hygrograph one end of a bundle of hairs is fixed and the other end moves a pointer that draws a line on a chart driven by a clock.

An instrument called a *psychrometer* consists of two thermometers. One of them, the dry-bulb thermometer, measures the air temperature; the other thermometer has its bulb covered with a jacket of muslin. This cloth is saturated with distilled water and then both thermometers are ventilated. In one version of this instrument, known as a *sling psychrometer*, the thermometers are mounted on a frame connected to a handle by means of a bearing or a small length of chain (Fig. 5-12). This psychrometer is whirled by hand in order to ventilate the thermometers. Some psychrometers have the thermometers fixed and the ventilation is supplied by a fan.

The procedure for using a psychrometer is to wet the muslin, ventilate the thermometers and read the dry- and wet-bulb temperatures. The former will soon read air temperature and remain constant with time. On the other hand, because of the cooling effects of evaporation, the temperature of the wet-bulb thermometer will decrease to a minimum known as the *wet-bulb temperature*. When the sling psychrometer is used, the practice is to make a series of swings and readings until the minimum temperature is established.

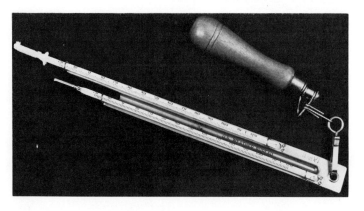

Figure 5-12 A sling psychrometer used for measuring air temperature and wet-bulb temperature from which humidity can be obtained.

The lower the humidity, the lower the wet-bulb temperature. If you know the wet- and dry-bulb temperatures and the atmospheric pressure, you can use a set of published tables to obtain the relative humidity of the air. One such table is given in Table A, Appendix II.

On a dry day the wet-bulb temperature can be much lower than the air temperature. For example, in Tucson, Arizona, it is not unusual in June for the maximum temperature to be 100°F (37.8°C) and the relative humidity to be 12 percent. As shown in Table A, Appendix II, at such a time the wet-bulb depression would be about 35°F and the wet-bulb temperature would be about 65°F (18.3°C). This fact accounts for the widespread use of evaporative coolers in houses and other buildings in hot, arid regions. These devices blow air over wet straw and produce temperature drops that on hot, dry days can amount to more than 20°C. Of course, when the relative humidities are high, evaporative coolers are little more than fans circulating the air.

If you know the dry- and wet-bulb temperatures, it is possible to obtain the *dew-point temperature* (see Table B, Appendix II). It is the temperature to which air must be cooled in order for saturation to occur when atmospheric pressure and water vapor content are held constant. When you see dewdrops on surfaces after a cool, moist night, it is a sign that the surfaces of the grass, car tops, etc., fell to the dew-point temperature during the night. One may obtain a measure of atmospheric moisture by means of a *dew-point hygrometer*. It consists of a polished, cooled surface whose temperature can be controlled and measured. As air moves over the surface, its temperature is reduced until droplets just begin to condense on the surface. The temperature at which this occurs is the *dew point*. If you know its value, the air temperature, and the pressure, you can calculate the wet-bulb temperature, the relative humidity, and the other humidity values mentioned earlier.

Another class of hygrometers makes use of the fact that certain substances change their electrical properties when exposed to varying humidities. The *radiosonde*, the balloon-borne instrument used to sound the atmosphere, employs a polystyrene slide coated with a thin layer of a mixture of carbon particles and a hydroxyethyl cellulose binder. Two electrodes on each side of the slide measure the electrical resistivity of the coating that increases as the relative humidity increases.

One could cite other procedures for measuring atmospheric humidity, but these are the most common ones. None of them is as accurate as a standard chemical technique that involves passing air through a substance that absorbs all the water molecules. By weighing the substance, you obtain the water vapor content of the air. Unfortunately, this procedure is time-consuming. Meteorologists find it more practical to use the much faster methods which pay the price of reduced but acceptable accuracy.

Let us now return to an examination of the change of humidity of a rising parcel of moist air. Consider the following example. A volume of air, close to sea level where the pressure is 1000 mb, has a temperature of 20°C and relative humidity of 50 percent. Figure 5-11 and Table 5-4 show that at this temperature the saturation vapor pressure is 23 mb, but since the relative humidity is 50 percent, the actual vapor pressure of the air at sea level is 11.5 mb. All these values are shown plotted, in Fig. 5-13, along the horizontal line labeled 1000 mb

Imagine that air at the ground begins to rise. As shown on the left side of Fig. 5-13, as the air ascends, it cools at the adiabatic rate, 10°C/km. The right-hand graph shows changes in the actual and saturation vapor pressures. Although the number of water vapor molecules in the rising parcel of air does not change, the total pressure decreases as the parcel expands. The water vapor density decreases and this accompanies a decrease of vapor pressure, as shown by the solid line. As the temperature decreases, there is a decrease in the saturation vapor pressure shown by the solid line (in accordance with the curve in Fig. 5-11). Since the saturation vapor pressure decreases more rapidly than does the actual vapor pressure, the curves converge. This means that as the air rises, its relative humidity increases, and at the altitude where the two lines intersect the actual and saturation vapor pressures are equal and the relative humidity is 100 percent. Continued ascent above this point, called the *condensation level*, would lead to a condition known as *supersaturation* where the actual vapor pressure exceeds the saturation vapor pressure. In general, before this occurs water molecules condense on tiny particles to form cloud droplets. The details of the formation of clouds will be discussed in a later chapter.

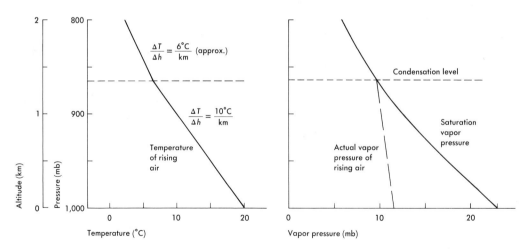

Figure 5-13   As a parcel of air rises, its temperature decreases at the dry adiabatic rate (10°C/km) until condensation begins at an altitude of 1.3 km. The actual vapor pressure and the saturation vapor pressure of the rising air decrease as shown. In this example the air at the ground, where the atmospheric pressure is 1000 mb, has a temperature of 20°C and a relative humidity of 50 percent corresponding to an actual vapor pressure of about 11.5 mb.

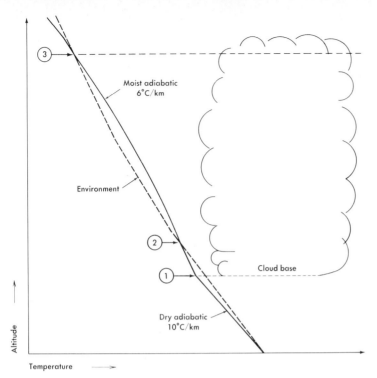

Figure 5-14 A rising volume of moist air cools adiabatically until condensation begins and a cloud forms. As the moist air continues to rise, it cools at the moist adiabatic lapse rate of about 6°C/km. Upward acceleration continues until the cloud air becomes cooler than the environmental air. The cloud tends to overshoot this level because of its upward momentum. (In order not to complicate this illustration, the warming effects resulting from freezing have been disregarded.)

At this time the essential points are that adiabatic cooling of ascending air leads to an increase of relative humidity, and if the air rises high enough, saturation and condensation can occur. When the air is very humid, an ascent of only a few hundred meters can cause cloud formation. If the air is extremely dry, it might have to rise many kilometers in order to become saturated. As can be seen from Fig. 5-13, if the air does not ascend enough, saturation would not be reached and there would be no condensation of cloud droplets. For example, such would be the case if the air ascended only to the level of 900 mb.

The condensation process not only leads to a conversion of water vapor to liquid; it also causes the release of latent heat. As mentioned in Chapter 3, the latent heat of vaporization varies with temperature, but it can be taken to be about 600 cal for each gram of water condensed. When you consider the huge quantities of water condensed in a storm system, it becomes evident that the latent heat represents a tremendous quantity of energy. A moderate rainfall, amounting to a fall of 6 mm, over a city the size of Washington, D.C., can release the energy equivalent to 20,000 tons of TNT.

The latent heat of condensation in a rising volume of air acts to partially compensate for the cooling caused by adiabatic expansion. The result is that once condensation begins, ascending air cools at a rate that is less than 10°C/km (Fig. 5-14). The new cooling rate is called the *moist adiabatic lapse rate*. It varies with temperature and pressure, but in the lower part of the atmosphere it is about 6°C/km.

When air containing water droplets descends, for example, in a thunderstorm downdraft, it warms at the moist adiabatic rate because the evaporation of water requires latent heat. The loss of about 600 cal for each gram of evaporated water reduces the compressional warming rate from about 10°C/km to 6°C/km of descent.

When rising air passes above the level where the temperature is 0°C, another source of heat becomes available, the *latent heat of fusion*. It is released when water freezes and amounts to about 80 cal/g. It is much smaller than the heat of vaporization, but it is nevertheless an important quantity in increasing the buoyancy of clouds, particularly of thunderstorms.

The effects of adiabatic cooling on a rising volume of moist air are summarized in graphical form in Fig. 5-14. In this example the low-level air is stable because the environmental lapse rate is smaller than the dry adiabatic rate. If a volume of air is forced, perhaps by a mountain barrier, to ascend from the ground to point 1, it becomes progressively cooler and more dense *in relation to the surrounding air*. At level 1, where the air reaches saturation, condensation begins and the rising air cools at the moist adiabatic rate. Between levels 1 and 2 the ascending air continues to have a downward buoyancy force exerted on it, and it still is necessary for a "mechanical force" to be applied in order for it to rise. Just above level 2, however, the ascending parcel of air finds itself in an unstable environment; it is warmer and less dense than the surrounding air. The resulting upward buoyancy force causes an upward acceleration of the cloud air. Continued convection leads to the formation of a towering convective cloud containing strong updrafts. When the ascending cloud air passes through level 3, it again finds itself colder and heavier than the environment air at the same altitude. The negative bouyancy means that there is a downward force that can rapidly decelerate the updraft and prevent any further cloud growth.

Up to this point it has been assumed that there is no mixing of cloud air and environmental air through the sides of convective clouds. In fact, outside air does pass through the boundaries of such clouds. This process, known as *entrainment*, has important effects on cloud temperatures and water contents.

In a positively buoyant, growing cloud the environmental air is cooler and drier than the cloud air at the same altitude. When the outside air is entrained, it causes temperature reductions in two ways. First, there is the addition of cool air. Second, some of the water particles in the cloud evaporate in order to saturate the drier air entering the cloud. The lowering of cloud temperatures reduces cloud buoyancy and inhibits cloud growth. The evaporation causes a reduction of the cloud's liquid water content. When clouds grow into a layer of very dry air, entrainment can lead to rapid desiccation and cloud dissipation.

The more humid the environment, the smaller are the consequences of entrainment. When the atmosphere is unstable and reasonably moist, an ascending current of buoyant air can be sufficiently broad that air mixed through the boundaries of a convective cloud does not reach the core of the updraft. In such a case, a towering thunderstorm can develop. The nature of such storms will be discussed in Chapter 9.

CHAPTER 6

Planetary Patterns of Air Motion

As Noted in Chapter 3, the atmosphere is sometimes called a heat engine because it is a system that receives energy in the form of heat, converts some of it to kinetic energy, and does work. Only a small fraction of the incoming solar radiation is transformed into the kinetic energy of air motions. Of the roughly 10^{14} kW of power received from the sun, about 2×10^{12} kW are converted into kinetic energy. The atmospheric engine therefore has an efficiency of only about 2 percent. This makes it a very inefficient engine, but nevertheless the available supply of kinetic energy in the winds is still enough to dwarf all man-made power sources.

When observing the wind fields in the earth's atmosphere, it is important to recognize that there are many scales of motion. For example, on the lowest scale there are gusts of wind having a dimension of centimeters and a duration of seconds. Swirling dust devils over the deserts have diameters of meters and lifetimes of minutes. Squall winds under thunderstorms cover areas that can be kilometers across and may last for hours. A line of thunderstorms can extend for distances of tens to hundreds of kilometers and go on for a day or so. Cyclones and anticyclones have dimensions from hundreds to thousands of kilometers and last for days.

At the largest scale of atmospheric motion are the planetary wave patterns extending over a major fraction of the entire globe. They are part of the general circulation of the atmosphere.

In order to understand patterns of weather and climate, it is necessary to appreciate the characteristics of the general circulation, the factors controlling it, and its interaction with smaller-scale circulations.

GENERATION CIRCULATION

If you examine a series of weather maps that show pressure and wind patterns over the earth, you find them changing continuously as pressures rise in some regions and fall in others. Centers of low and high pressure weaken or strengthen markedly over half a day. Over periods of a few days old pressure centers disappear and new ones form. A casual examination might give the impression of a random series of unrelated events in time and position. Such an idea, however, is not in accordance with the behavior of the atmosphere. Certain important features of the general circulation, because of their persistent nature, show up clearly when weather maps for many years are averaged.

An idealized version of the wind patterns over the earth is shown in Fig. 6-1. Equatorward from the belts of high pressure at latitudes of about 30°N and 30°S, one finds the *trade winds* from the northeast on the north side of the equator and from the southeast on the south side of the equator. In middle latitudes there are prevailing westerly winds; another belt of easterlies is found near the poles. Average surface pressures and winds for winter and summer are shown in Fig. 6-2.

As is illustrated in Fig. 6-1, in regions of low pressure the air tends to rise. It is not surprising to find that in equatorial regions, where the trade winds converge, the air generally ascends. Since it is often very humid, the result is heavy rainfall. The regions where the trades meet is called the *intertropical convergence zone* (ITCZ) or the *equatorial through*. It is a low pressure area where surface winds generally are light and for this reason it was named the *doldrums* in the days of the great sailing ships. On the average,

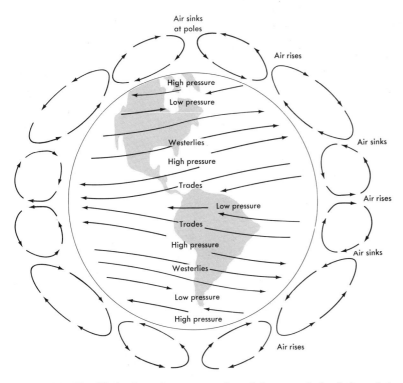

Figure 6-1 Simplified schematic representation of the general circulation of the atmosphere.

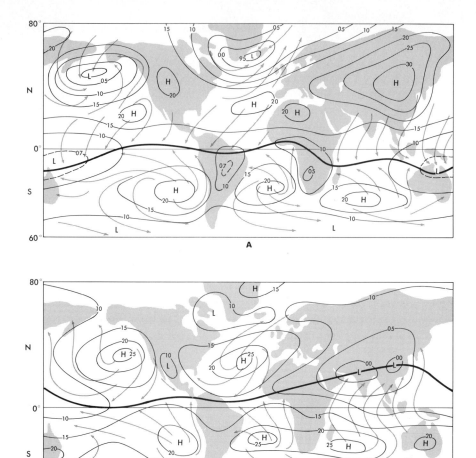

Figure 6-2 Average sea-level winds and pressures (isobars in millibars exceeding 1000 mb) over the earth in (A) January and (B) July. The heavy solid line is the intertropical convergence zone. From *Introduction to the Atmosphere* by Herbert Riehl. Copyright 1972. Used with permission of McGraw-Hill Book Company.

the air ascends over the ITCZ, moves poleward, and descends in the regions of higher pressure in the subtropics. This circulation resembles a giant convection cell. It is some times called the Hadley cell, after George Hadley who discovered it early in the eighteenth century.

The most persistent features of the general circulation are the high-pressure cells over the oceans (see Fig. 6-2). They are sometimes described as being "semipermanent" and are given specific names. The Pacific anticyclone or Pacific high is usually centered between 30°N and 40°N latitude and 140°W to 150°W longitude. It exerts major control over the climate of the west coast of the United States and Mexico. The dry subsiding air explains the presence of the Sonora desert of Mexico and Arizona.

Over the Atlantic ocean the high-pressure region is called the Azores high or the Bermuda high according to the longitude of the middle of the anticyclone. This giant swirl of high pressure dominates the climate of southern Europe and North Africa. The subsiding air over these places accounts for the dryness of the Mediterranean countries and existence of the Sahara desert.

The semipermanent anticyclones in the Southern Hemisphere exert major controls over the weather and climate in Africa, Australia, and South America. Persistent subsidence along the eastern branches of the high-pressure cells is the principal reason for the existence of the Namib desert of Southwest Africa, the Peru and Atacama deserts along the coastal areas of Peru and Chile and the desert areas of western Australia. The cold ocean currents along the west coasts of North and South America and Africa act to stabilize the air, inhibit cloud formation, and thereby contribute to desert formation.

Near the center of the high-pressure areas the winds are light. In the days of the sailing ships, vessels sometimes were becalmed for long periods. This region was named the *horse latitudes*, presumably because horses had to be dropped over the side when the supply of food and water ran out.

It is seen in Fig. 6-2 that the patterns of pressure over the earth change markedly from winter to summer. This is a reflection of the differences of air temperature from season to season, as is shown in Fig. 3-12. During the winter air temperatures are colder over the land than over the oceans. As a result, atmospheric pressures generally are higher over the continents. This is particularly evident over massive Eurasia. In the winter the subtropical highs over the oceans are relatively small, but the low-pressure systems that prevail over the Aleutian Islands and Iceland are well developed.

In the summer air temperatures over the land generally are higher than over the water. As a consequence, there is substantially higher pressure over the oceans than over the continents. The Pacific and Azores highs are large and persistent, and the formation of low-pressure centers over the northern parts of the oceans is inhibited. Over the continents there tends to be low pressure, particularly at the lower latitudes—over Mexico and India, for example.

The changes of pressure over continents from winter to summer and particularly the changes in the pressure gradient between land and water are very important. They determine the direction of the air currents and explain the occurrence of the monsoon circulations mentioned in Chapter 4. In the winter, as shown in Fig. 6-2, Asia is under the influence of a strong, persistent high-pressure region. The consequence is a prevailing outward flow of air across southeast Asia over the Indian Ocean and the South China Sea. This is the *winter monsoon*, consisting of cool, dry air with little if any precipitation.

In the summer the continent warms and, as noted above, southeast Asia is dominated by low pressure. As a consequence, the *summer monsoon* becomes established and warm, humid air sweeps northward over the conti-

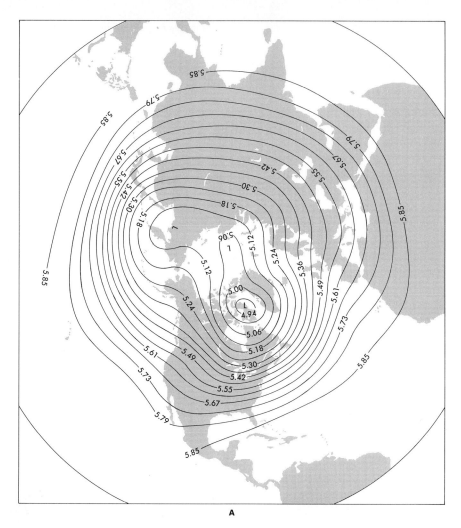

Figure 6-3 Average 500-mb charts over the Northern Hemisphere in (A, *above*) January and (B) July. The heavy lines show the height of the 500-mb level in kilometers. *From* Technical Report 21, U.S. Department of Commerce, National Oceanic and Atmospheric Administration.

nent and is lifted by the high terrain. Showers and thunderstorms form in abundance and yield the heavy rains that are so crucial to the farmers of that part of the world. Most of the rains are a result of weak cyclones, called *monsoon depressions*, which propagate westward across southeast Asia. The climatic variations in the southern United States, particularly the southwest, are also monsoonal in character, but they are not nearly as pronounced as in southeast Asia. Monsoon circulations also are observed in other parts of the world.

The variations of pressure with season of the year are much more pronounced in the Northern Hemisphere than in the Southern Hemisphere

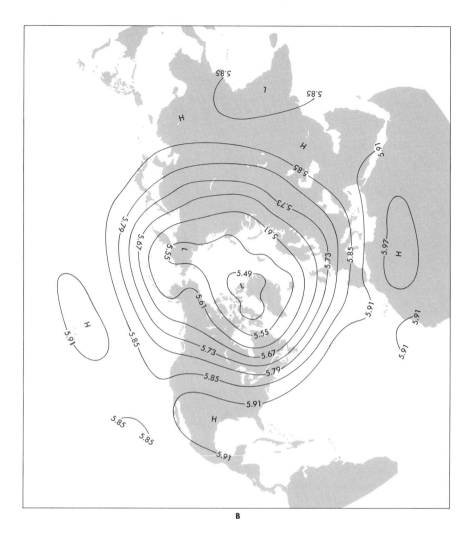

Figure 6-3B

because north of the equator there are extensive continents. About 40 percent of the area is occupied by land. The Southern Hemisphere is mostly water and the oceans exert major control over the hemispheric temperature and pressure distributions.

The maps in Fig. 6-2 show how the patterns shift with the seasons. The semipermanent centers of high pressure are farther north in July than in January. It can be seen that the intertropical convergence zone, represented by the dark, wavy line, "follows the sun." In January its average position is at about 5°S latitude while in July it averages about 10°N latitude.

The patterns of pressure and wind are smoother in the upper atmosphere than at the surface. Figure 6-3 shows average conditions at the 500-mb levels in the Northern Hemisphere. In the winter there are two low-pressure centers located generally to the west of their surface counterparts over the

Aleutians and Iceland. Except for these distinct centers, you see mostly a broad current of westerly winds blowing through long-wave patterns. Over the United States, western Europe, and the western Pacific, the lines of constant height dip toward the south. They are called *long-wave troughs* and are separated by *long-wave ridges*.

Because of the relatively large temperature difference between tropical and polar regions in the winter, there are large north–south pressure gradients and strong westerly winds. The overall latitudinal temperature and pressure gradients are smaller in the summer and therefore the westerlies are weaker, as shown in Fig. 6-3(B). It also shows a belt of high pressure at low latitudes. To the south of this belt there are easterly winds at the 500-mb level. Some low-latitude easterlies are also found in the winter, but they usually are nearer the equator than they are in the summer.

JET STREAMS

One of the striking features of the circulation pattern in the earth's atmosphere becomes dramatically evident when you examine vertical cross sections through the atmosphere along a north–south line. Such a cross section is shown in Fig. 6-4. It displays the distribution of temperature through a well-developed cold front that extends into the stratosphere. The nature of fronts will be discussed in the next chapter. At this point we shall ignore the presence of the front and focus on the temperature and wind structure. The cross section shows that there is a concentrated current of high-speed air centered at an altitude of about 10 km. Its maximum speed exceeds 80 m/s (179 mi/hr). Such a strong wind current is called a *jet stream*.

The *polar-front jet* is observed at middle latitudes, at an average altitude of about 12 km and exhibits maximum speeds averaging about 60 m/s (134 mi/hr). In extreme cases, it is more than twice as fast. On some occasions this jet stream extends around almost the entire globe. As shown in Fig. 6-5, the average position of the jet meanders northward and southward, and it exhibits maxima and minima of wind speed along its path. Because of the variations from day to day in position and speed, average maps such as the 500-mb charts in Fig. 6-3 do not show the tight pressure gradient you would expect with a jet such as the one shown in Fig. 6-5. Nevertheless, it is seen from the closeness of the lines of constant height in Fig. 6-3(A), that strong westerly winds exist at middle latitudes.

Detailed analyses of three-dimensional air flows over the Northern Hemisphere reveal that there are several tropospheric jet streams and that they vary with season of the year. As already noted the polar front jet generally exhibits high speeds. The westerly-blowing subtropical jet stream, on the average, is centered at a latitude of 25°N, at an altitude of 13 km and has a speed of about 40 m/s. In its equatorward meanderings the polar front jet

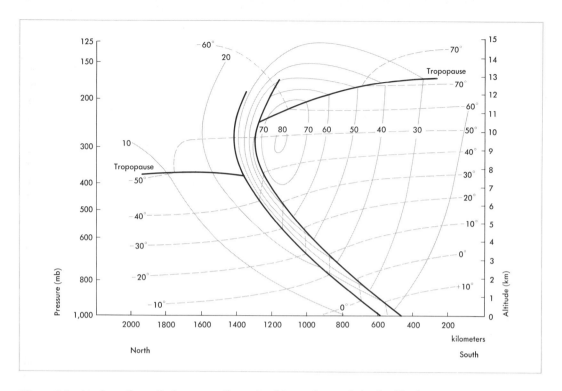

Figure 6-4 A schematic vertical cross section extending north–sourth in the Northern Hemisphere through the polar front (heavy lines) and showing the wind profile through a jet stream. The speeds are in meters per second and temperatures are in degrees Celsius. *From* E. Palmén and C. W. Newton, *Atmospheric Circulation Systems*, Academic Press, 1969, based on analyses by R. Berggren.

sometimes merges with the subtropical jet stream. In the summer at low latitudes, a relatively weak easterly jet sometimes is observed at altitudes of about 16 km, mainly over southeast Asia.

A jet stream exists because the north–south pressure gradient increases with height up to the level of the jet. As suggested by Fig. 6-4, this occurs because of the large north–south temperature gradients under the jet. On the south side the air is warmer than on the north side. In accordance with the hydrostatic equation discussed in Chapter 2, the pressure decreases with height more rapidly the lower the temperature. For this reason, pressure decreases more rapidly with height on the north side of the jet than it does on the south side. This explains the increase of pressure gradient with height below the jet. Above it, as shown in Fig. 6-4, there is a reversal of the north–south temperature gradient with cold air to the south and warm air to the north. With increasing height, the north–south pressure gradient diminishes and so does the wind speed.

In general, the middle-latitude jet is stronger in the winter than in the summer because there are greater temperature differences with latitude. In January the temperature difference between equatorial and Arctic zones can

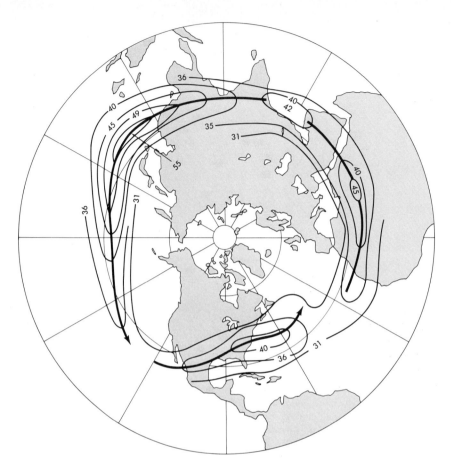

Figure 6-5 The average position and strength of middle-latitude jet stream in January.
From J. Namias and P. F. Clapp, *Journal of Meteorology*, 1949, **6**: 330–336.

amount to about 70°C, while in July it is about half that much. As illustrated in Fig. 6-4, there is not a uniform decrease of temperature with latitude. Instead, there is generally warm air to the south separated from colder air to the north. The transition region between them is called the *polar front* and the associated strong wind current is the polar-front jet. There are very pronounced changes of temperature in the north–south direction through the polar front. That accounts for the presence of the jet stream just above the polar front.

In the winter, cold air pushes farther southward than it does in the summer. As a result, the average position of the jet stream is also closer to the equator in winter than in summer.

The jet stream plays an important role in governing the behavior of the atmosphere. It is a means for the rapid propagation of energy over long distances. In discussing Fig. 6-3 we pointed out the average positions of long-wave troughs at 500 mb. These waves and smaller, but still important, low- and high-pressure perturbation move and change in amplitude. The high velocities in the jet stream can propagate the effects of pressure disturbances around the globe. Winds in the jet core sometimes are high enough to carry air, at latitudes of 40°N to 50°N, around the earth in five days. The jet stream winds can bring about the development of low-level cyclones and anticyclones.

Detailed analyses of the three-dimensional characteristics of jets show that the wind speeds through them are not constant along the jet core. In some segments the wind speeds are higher than in other segments. Air accelerates as it moves into the regions of maximum speeds and causes air mass divergence. This divergence aloft leads to upward air motions, a fall of pressure at the surface, and the initiation of cyclonic development. As air leaves a high-speed segment, there is air mass convergence at jet altitudes. It leads to higher pressures and subsiding air.

The steady component of the strong westerly winds in the jet stream also have associated with it vertical air motions having important meteorological and practical consequences. Figure 6-6 is a schematic representation of the essential features of a polar-front jet stream. Note, in particular, that there is a stream of air from the stratosphere down through the frontal zone into the lower atmosphere. This is the principal means by which gaseous and particulate substances from the stratosphere are brought toward the ground. For example, the sinking air removes ozone from the stratosphere. Radioactive particles put there by atmospheric nuclear tests are carried to the lower atmospheres through the jet stream gap in the tropopause. Minute particles from volcanic eruptions also are swept out of the stratosphere by the same route. Once into the troposphere, the particles are washed out by rain and snow. It is not surprising that the highest concentrations of so-called *stratospheric debris* falls over the middle latitudes where the polar jet resides most often.

Before leaving the middle-latitude jet it is necessary to mention its importance to the aviation industry. High-flying airplanes going from west to east can save a great deal of time and fuel by flying in the high winds of the jet stream. An airplane normally cruising at an air speed of 260 m/s (582 mi/hr) might take about 7 hours to fly from New York to Paris if the average current wind speed were 25 m/s (56 mi/hr). On the other hand, if it flies in a jet stream averaging 75 m/s (168 mi/hr), the trip to Paris would take about 6 hours. Of course, on the flight toward the west, an airplane would seek to avoid the jet stream. Airline meteorologists make use of a knowledge of high-level wind data to plan flight tracks.

A problem arising in flights along or through the jet stream is the presence of turbulence. Because it frequently occurs with cloud-free skies, it is called *clear air turbulence* or CAT for short. The strong wind shears cause highly variable patterns of up and downdrafts over distances on the order of the size of an airplane. The result can be violent buffeting as severe

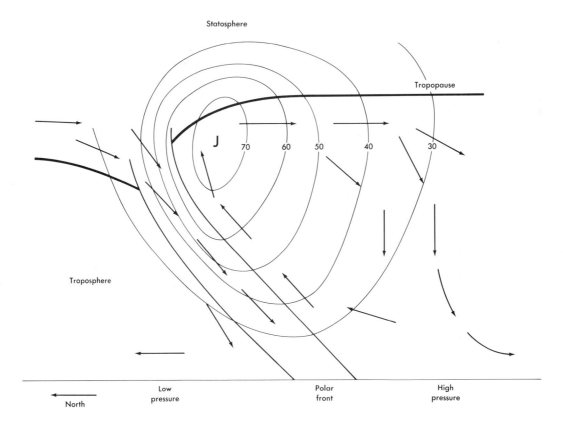

Figure 6-6   Schematic version of Fig. 6-4 showing the pattern of air velocities in a vertical plane through the middle-latitude jet stream. It is sometimes called the *polar-front jet*.

as would be expected in a severe thunderstorm. Jet stream CAT is variable and its precise location difficult to predict and, at this time, impossible to detect until an airplane is in it. The usual remedial action is to change altitudes or horizontal track in an attempt to find smoother air.

As noted earlier, the subtropical jet stream located at low latitudes in the Northern Hemisphere, blows from the west in the winter. As summer approaches, the westerlies are replaced by easterlies. Sometimes they are strong enough to constitute the *easterly jet stream*. It effects the lives of millions of people in a crucial way. The change in direction of the jet stream is most dramatic over the Asian subcontinent where it can occur over a period of a few days. The westerly jet, which appears in late September, is characterized by sinking air and fair weather. In normal years, toward the end of May, the west wind jet moves northward over the Himalayas and merges into the polar-front jet. Over India and other parts of southern Asia the easterly subtropical jet becomes established. It marks the onset of the summer monsoon as warm, humid air from the Indian Ocean moves over India and neighboring countries.

INTERACTIONS
OF AIR AND SEA

For reasons discussed in Chapter 3, the oceans have been called the thermostat of the earth. They represent a huge fluid mass having a large heat capacity. They store an enormous quantity of energy and exchange it readily with the atmosphere. In winter the oceans serve to warm the air moving over them. (See the maps in Fig. 3-12.) In summer the ocean water tends to be cooler than the surface air, and hence it represents a massive sink for heat transferred from the air.

The maps in Fig. 6-2 show the effects of land and ocean temperature differences on the surface pressure patterns over the earth. These patterns, in turn, determine systems of weather and wind flow.

The oceans are crucial not only as sources and sinks for heat, but also, as noted in Chapter 3, as the means by which large quantities of energy are transported from the warm equatorial regions to the colder polar regions. In the process, they reduce the overall temperature difference and hence the driving force in the global wind system. The warm ocean currents such as the Gulf Stream and the Kuroshio and Brazil Currents transport heat poleward. The California and Peru Currents carry cold water equatorward (see Fig. 3-14.)

The ocean bodies of the world also affect the general circulation in ways which, in the past, have received little attention. One of these is associated with the upwelling of large quantities of cold water in the equatorial regions. This occurs in response to the deflection of surface water in opposite directions by the northeast and southeast trade winds. It has been reported that in the eastern Pacific the upwelling water, which is several degrees colder than the surrounding water, sometimes extends over regions several thousand kilometers across. During certain years there is no upwelling of cold water, but during other years bodies of low-temperature water suddenly appear. Jakob Bjerknes, while at U.C.L.A., reported that the occurrence of large tongues of upwelling water was related to changes in the atmospheric circulation of the Northern Hemisphere.

Another prominent meteorologist, Jerome Namias, has observed that during some years water temperatures in the North Pacific are as much as 6°C above the long-term average. It has been speculated that such anomalies might exert some control over the general circulation of the atmosphere.

The degree to which the phenomena of sporadic, large-scale upwelling or abnormal warming occur over the earth's oceans still needs to be ascertained. Earth-orbiting satellites equipped with appropriate radiometers facilitate the detection and measurement of anomalies in ocean temperatures.

The interaction between sea and air can also be greatly influenced by the formation of sea ice. Joseph O. Fletcher, who has pioneered investigations of this subject, has noted that ice is a good insulator. In the Arctic less than a meter of ice can maintain a temperature of the ice surface at $-30°C$ while the ice is in contact with ocean water at $-2°C$. The ice effectively reduces the transfer of heat from water to air.

A second important role of ice is to increase the reflectivity of the surface. In summer open polar oceans absorb about 90 percent of the insolation. This compares with the 30 percent to 40 percent presently absorbed by the highly reflective year-round ice.

The effect of the sea ice is therefore to suppress heat transfer from ocean to air and to reduce the quantity of absorbed solar energy. An increase in ice cover amplifies these effects leading to colder temperatures, more sea ice, and an extension of the same process. This tendency for a process to sustain or amplify itself is sometimes referred to as *positive feedback*. Once started, it can continue until some other outside mechanism comes into play and reverses the process. Once the reverse process begins, it will continue once again because of its positive feedback aspects.

How far the ice extension or elimination process must proceed before it significantly influences the general circulation of the atmosphere still is not known. Records show that over past years there have been great changes in the extent of the sea ice. For example, for about half a century ending in about 1940 there was a gradual warming of the earth accompanied by a reduction of the equatorward extension and thickness of the sea ice. As shown in Fig. 11-16, about 1940, for reasons which still are not clear, the earth began to cool and the process of sea-ice formation reversed. Apparently the global cooling ended in the early 1960's.

The Soviet climatologist, M. I. Budyko, predicted that if the Arctic pack ice were totally melted, the present incoming solar radiation would prevent it from reforming. Instead, he visualized an ice-free Arctic Ocean and a different climatological region in the Arctic. One would also expect that the temperature difference between equatorial and polar latitudes would be smaller than at present and would lead to major changes in the general circulation of the entire atmosphere. There still is considerable uncertainty about Budyko's hypothesis of an ice-free Arctic and the effects it would have on global circulation.

If the warming in the higher latitudes had continued beyond 1940 and gone on long enough to produce changes in deep ocean temperature, nature might have tested Budyko's hypothesis. Instead, there was a period of general worldwide cooling for the subsequent two decades. As a result, a check of the hypothesis can only be made by means of theoretical models of the general circulation that realistically account for atmospheric-ocean interactions.

LABORATORY MODELING
OF THE GENERAL CIRCULATION

The character of the general circulation depends on various properties of the earth and the atmosphere. A very important factor is the temperature difference between equatorial and polar regions. Other critical elements are the rotational speed of the earth, the configurations of land and sea, and the physical and chemical properties of the air.

Certain aspects of the general circulation can be modeled with impressive success in laboratory experiments. They were first performed by Dave Fultz at the University of Chicago in the late 1940's. The technique consists of using water to represent the air and a circular pan to represent the earth. The pan is heated along the outside rim and cooled at the center where a circular cylinder is fastened. In this way the experiment simulates equatorial heating and polar cooling. The pan is rotated at varying speeds to simulate the earth's rotation. Many other experimental variations can be introduced, for example, obstacles can be used to simulate the effects of mountains. Dyes can be injected into the liquid to serve as markers for tracing fluid motions. Tiny probes can be used to measure properties of the fluid.

Laboratory studies such as these have yielded many interesting and important results. They verified a discovery made earlier by the famous meteorologist, Carl-Gustav Rossby, that the character of fluid flow on a rotating body depends on the ratio of the characteristic velocity of the fluid to the characteristic velocity of the body. This quantity, known as the *Rossby number*, for the earth is about 0.1, i.e., the ratio of the speed of the jet stream (about 50 m/s) to the speed of the earth surface at the equator (about 500 m/s). When the dishpan experiment was performed at a Rossby number of about 0.1, the pattern of fluid motion (Fig. 6-7) was similar in many respects to the wind pattern in the free atmosphere.

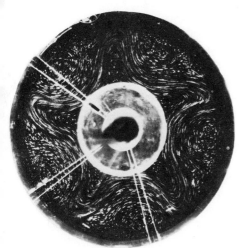

Figure 6-7 Laboratory simulation of general circulation of the atmosphere. *Courtesy* Dave Fultz, Hydrodynamics Laboratory, University of Chicago.

As might be expected, the fact that the dishpan is flat while the earth is nearly spherical presents some difficulties. At the poles the earth's surface is perpendicular to the axis of rotation as is the surface of the dishpan. As one goes toward the equator, the earth's surface becomes more and more parallel to the axis, thereby reducing the effects of the rotation. This means that the pattern of flow in the real atmosphere at low latitudes has a larger Rossby number than 0.1. It also suggests that the rotation in the tropical atmosphere should be simulated by means of dishpan experiments at slow rotational speeds. This relation between latitude and speed was borne out by experiment.

These results indicate that the character of the general circulation, and particularly the separation of the tropical and middle-latitude atmosphere, is closely coupled with the rate of rotation of the earth. Herbert Riehl, an authority on tropical meteorology, speculated that if the earth's rotation were one-half to one-quarter of the present rate, the low-latitude convective circulation, which as shown in Fig. 6-1 now extends to latitudes of about 30°, might reach as far as latitude 60°. Such considerations are of crucial importance in developing models of the atmospheric motions of other planets that rotate at speeds far different from those of the earth.

THEORETICAL MODELS
OF THE GENERAL CIRCULATION

Although laboratory models of the general circulation. have yielded new information, they have some obvious limitations associated with difficulties in accurately simulating various important processes. The dishpan experiments do not include energy transfer processes such as those associated with radiation and cloud and precipitation formation. Also, the effects of the ocean and of the snow- and ice-covered regions of the earth, sometimes referred to as the *cryosphere*, are not taken into account.

In the view of many atmospheric scientists, the most promising approach to the study of the general circulation is through the medium of mathematical models. This is not a new idea, but it was totally impractical until the development of large, high-speed electronic computers.

A mathematical model of the general circulation starts out with a number of equations that specify the nature of the system and how it changes with time. A typical model consists of the following parts:

1. An equation of state relating pressure, temperature, and density of the air.
2. An equation of motion relating changes with time in the three-dimensional air motions to the pressure, gravitational, and friction forces.
3. Thermodynamic equations dealing with temperature changes in the air–earth system.

4. Equations dealing with water vapor, clouds, and precipitation.
5. Equations dealing with the transfer of radiant energy through the atmosphere.
6. Equations dealing with heat, momentum, and water balance at the earth's surface.

The equations describing the general circulation are interrelated and must be solved by means of techniques known as *numerical methods*. They involve the specification of pressure, temperature, and humidity over a grid of points spaced perhaps 400 km apart horizontally. Conditions are specified at a number of pressure levels, sometimes as many as ten of them between 1000 mb and 10 mb. This includes the altitude range from about sea level to 30 km. Calculations are made of the changes of temperature, humidity, and pressure over a brief interval of time. This establishes a new set of initial conditions and a second calculation of changes is made. By means of a long series of time steps, calculations can be made of the evolution of the state of the atmosphere.

In practice, when the development of the general circulation is being calculated, simple conditions are assumed to prevail at the outset. For example, it is sometimes assumed that the atmosphere is at rest and is isothermal, i.e., it has the same temperature everywhere. Then the sun is "turned on" and all the processes incorporated in the mathematical model are allowed to function. During the first few "days" there is little motion as the equatorial regions get progressively warmer than the poles. When temperature differences become large enough, convection and three-dimensional air motions begin. The effects of the earth's rotation, sea–land differences, mountain ranges, cloud and rain formation, ice-covered surfaces, and other factors come into play. After about 200 to 300 model days, calculated general atmospheric circulations develop which have striking similarities to those actually observed (see Fig. 6-8).

The models cannot yet treat radiative transfers of energy in a satisfactory way. In particular, the effects of aerosols and certain trace gases still are not adequately known. To a certain extent, this is because of insufficient information about the quantities and characteristics of these substances.

Another major problem area for the modelers is the interaction of sea and air. The two fluid systems are closely coupled. A change of temperature of one medium can lead to a change in the other. Mathematical models have been developed that take into consideration exchanges of radiant heat and turbulent transport of sensible heat as well as latent heat of vaporization. Water exchanges as a result of evaporation from the oceans and precipitation into the ocean are built into the model. Finally, the mathematical equations take into account exchanges of momentum produced by wind stresses at the ocean surface. The resulting series of expressions still do not include all the exchange mechanisms. Nevertheless, a numerical model developed by Syukuro Manabe and Kirk Bryan at the Geophysical Fluid Dynamics

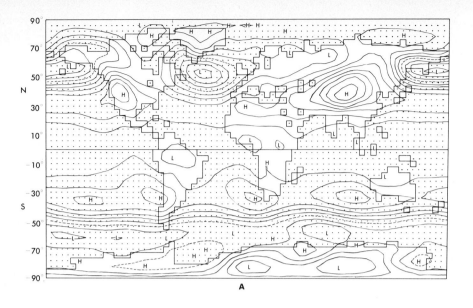

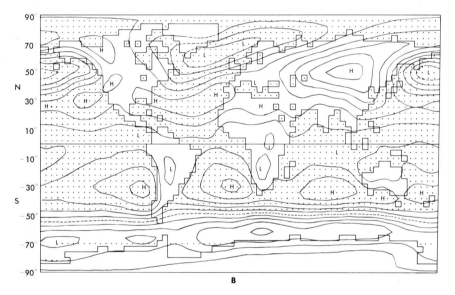

Figure 6-8 Computer-generated mean sea-level pressure in January: (A) computed and (B) observed. Isobars are at intervals of 4 mb and broken line is 1000 mb. *From* Y. Mintz, A. Katayama, and A. Arakawa, University of California at Los Angeles, 1972.

Laboratory of the National Oceanic and Atmospheric Administration yielded the encouraging results shown in Fig. 6-9. The model shows that the computed temperature pattern in the atmosphere and oceans is in reasonably good agreement with observed pattern.

A great deal of progress has been made in developing theoretical models of the general circulation of the atmosphere and the interaction of atmosphere and oceans. These models offer the hope of being able to make reliable weather forecasts one to two weeks in the future. If the state of the atmosphere at any time is known, the mathematical model is used to calculate changes over future time periods. For this purpose more complete observations are needed of the earth's atmosphere than are currently being obtained. There are large open spaces over the oceans, particularly over the Southern Hemisphere. Remote sensing instruments on earth-orbiting satellites are being used to collect some of the required measurements; instrumented aircraft, ships, and buoys also are augmenting the worldwide network of surface and radiosonde network.

The general circulation of the atmosphere, as already noted, depends on a great many factors. Some of them are external to the earth. For example, the character of the temperature, pressure, and wind structure depends on the nature of solar radiation and the orbit of the earth around the sun. The circulation depends in a crucial way on the geography of the planet and on the constituents of the atmosphere.

Once a realistic mathematical model of the general circulation of the atmosphere is developed, it will be possible to evaluate the effects of changes in both internal and external factors. One might ask, for example,

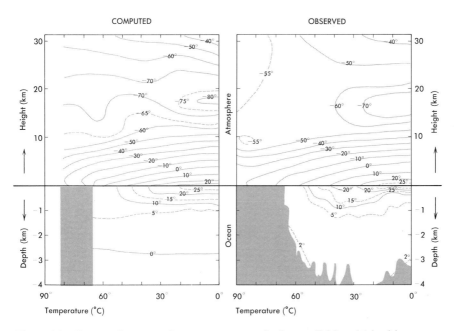

Figure 6-9 Computed pattern of mean temperature in degrees Celsius obtained by averaging the patterns calculated for both hemispheres around latitude zones. The right-hand side shows observed temperatures in the Northern Hemisphere atmosphere and in the North Atlantic Ocean. *From* S. Manabe and K. Bryan, *Journal of Atmospheric Sciences*, 1969, **26**: 786–689.

what would happen if the solar constant changed by 1 percent, or what is the effect of the increasing carbon dioxide, or the change in atmospheric turbidity caused by volcanic eruptions or excessive worldwide atmospheric pollution. Unfortunately, at this time the mathematical models of the general circulation are not sufficiently well developed to give more than tentative answers to these and to other questions like these. Major research efforts are in progress in various research establishments to refine the models for use in the study of climate and for predicting the weather one to two weeks in advance. Mathematical models already have been found to be very useful in forecasting the weather for periods of one to two days. This topic will be examined in Chapter 12.

CHAPTER 7

Air Masses, Fronts, and Cyclones

THE GENERAL CIRCULATION OF THE ATMOSPHERE can be said to describe average patterns of pressure and winds over the earth. As you would expect, at any particular time the configuration of isobars, winds, temperatures, and other weather features can differ substantially from the average.

The state of the atmosphere is monitored at regular intervals at weather stations all over the world. They measure temperature, pressure, wind velocity, and humidity near the ground and observe the state of the sky. They note the type and extent of clouds and such weather features as rain, snow, lightning, and thunder. At three-hour intervals the data, in coded form, are transmitted via teletype and radio to various central offices and are used for constructing synoptic charts showing sea-level weather conditions.

The weather data plotted on surface maps are arranged in a specific way. Figure 7-1 shows the so-called abbreviated U.S. model for displaying weather observations. Often some of the indicated quantities are omitted. The most essential items are wind direction and speed, pressure (ppp), pressure change (ppa), temperature (TT), dewpoint (T_dT_d), present weather (ww), and total amount of cloud cover. Wind direction is indicated by an arrow and wind speed by the number of barbs on the arrow. The types of clouds (C_H, C_M, C_L), the present weather (ww), and the past weather (W) are designated by symbols. There are 10 categories of high, middle, and low clouds and 100 categories of weather types. Table 7-1 shows some of the symbols most commonly used for indicating sky cover, cloud type, pressure change, and weather types.

Figure 7-2 shows a plotted and analyzed weather map of the United States. Normally there are more plotted stations. An examination of such a map reveals that conditions do not vary randomly from place to place. There are large regions where the temperatures and dew points are fairly uniform. As will be seen below, the boundary separating two fairly uniform masses of air is called a *front*.

Synoptic weather charts also show that atmospheric pressure varies in a fairly smooth fashion from place to place. When you draw isobars, centers of high and low pressure become evident. Often "the weather"—the clouds, rain, and snow—are associated with the fronts and regions of low pressure. High-pressure areas commonly exhibit fair weather.

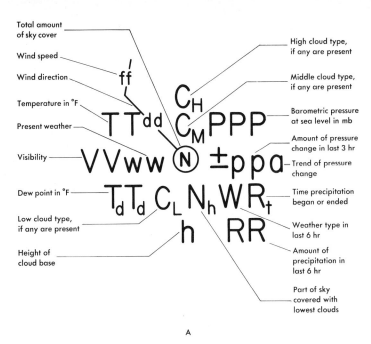

Total amount of sky cover

Wind speed

Wind direction

Temperature in °F

Present weather

Visibility

Dew point in °F

Low cloud type, if any are present

Height of cloud base

High cloud type, if any are present

Middle cloud type, if any are present

Barometric pressure at sea level in mb

Amount of pressure change in last 3 hr

Trend of pressure change

Time precipitation began or ended

Weather type in last 6 hr

Amount of precipitation in last 6 hr

Part of sky covered with lowest clouds

A

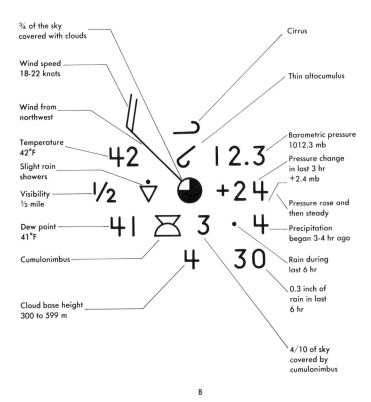

¾ of the sky covered with clouds

Wind speed 18-22 knots

Wind from northwest

Temperature 42°F

Slight rain showers

Visibility ½ mile

Dew point 41°F

Cumulonimbus

Cloud base height 300 to 599 m

Cirrus

Thin altocumulus

Barometric pressure 1012.3 mb

Pressure change in last 3 hr +2.4 mb

Pressure rose and then steady

Precipitation began 3-4 hr ago

Rain during last 6 hr

0.3 inch of rain in last 6 hr

4/10 of sky covered by cumulonimbus

B

Figure 7-1 (A) Station model used for plotting data on a surface or sea-level weather map. (B) Example of a plot.

Table 7-1 COMMON SYMBOLS USED ON WEATHER MAPS

Total amount of sky cover:

○	⊙	◔	◕	◑	⊕	◖	◗	●	⊕
0	1	2-3	4	5	6	7-8	9	10	

Each number is the number of tenths of the sky covered.
The last symbol means that the sky is obscured
(by dust or smoke, for example)

The primary cloud forms are the following:

╱ Cirrus

∠ Cirrostratus

⌇ Cirrocumulus

⌄⌄ Altocumulus

⌇ Thin altocumulus

∠ Altostratus

⟋ Nimbostratus

— Stratus

⌄ Stratocumulus

⌒ Cumulus

⌂ Cumulus congestus

⊟ Cumulonimbus

Pressure change over the last 3 hours (ppa):

╱ Steady rise

⌐ Unsteady rise

⌐ Unsteady fall

⟋⟍ Rise, followed by a larger fall

⟍⟋ Fall, followed by a lesser rise

Commonly used weather (ww or W) symbols:

∞ Haze

≡ Fog

⌐ Smoke

' Drizzle

• Rain

✱ Snow

△ Hail

▽ Shower

Ʀ Thunderstorm

Ʀ Heavy thunderstorm

Ʀ Thunderstorm with hail

< Lightning

⌁ Dust or sandstorm

⊬ Drifting snow

●⌄ Freezing rain

•⌐ Rain ended during preceding hour

Table 7-1 COMMON SYMBOLS USED ON WEATHER MAPS (cont.)

The intensity of drizzle, rain, or snow is indicated by the number of repetitions of the symbol. For example:

- • Light, intermittent rain

- •• Light, continuous rain

- ⦂ Moderate, intermittent rain

- ∴ Moderate, continuous rain

- ⦙ Heavy, intermittent rain

- ⦙• Heavy, continuous rain

Wind direction (dd) and speed (ff) are indicated by the direction of the arrow and the number of barbs:

⊙	Calm		38–43 mi/hr
——	1–2 mi/hr		44–49
	3–8		50–54
	9–14		55–60
	15–20		61–66
	21–25		67–71
	26–31		72–77
	32–37		78–83

Weather forecasters keep track of the pressure centers and fronts as they evolve and move. In order to forecast high and low temperatures and whether or not precipitation will occur, it is necessary to predict the character of future weather maps. For obvious reasons, they are known as prognostic charts.

AIR MASSES

Although air temperatures near the ground are highest in equatorial regions and lowest at the poles, it is well known that there is not a gradual decrease of temperature with latitude. Instead, at lower latitudes there is a widespread body of warm air whose temperature decreases only gradually with latitude. Over polar regions there commonly is a massive sea of air having fairly uniform properties and temperatures that increase slowly toward the equator. These widespread bodies of uniform air are called *air masses*.

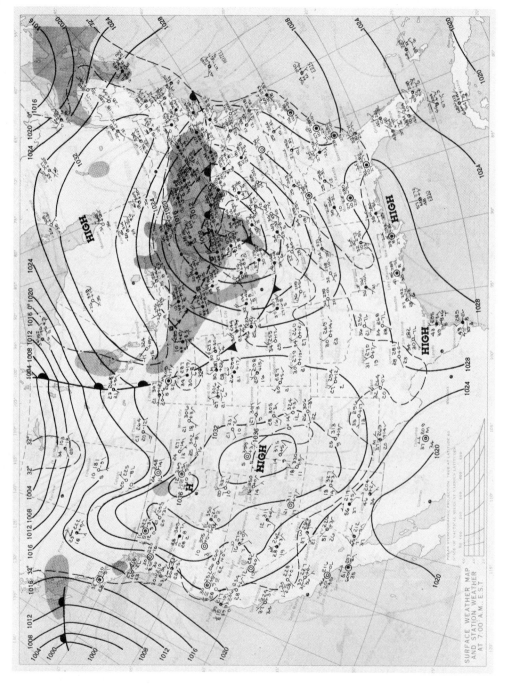

Figure 7-2 A plotted weather map showing isobars and fronts. Note that, usually, some of the data shown in the model in Fig. 7-1 are omitted. This is a more complete version of the map shown earlier in Fig. 4-6. The shaded areas are regions where snow was falling.

An air mass develops its characteristic properties by remaining over a particular region of the earth for periods long enough to allow its vertical distributions of temperature and moisture to reach equilibrium with the underlying surface. The processes by which this occurs will become evident as various air masses are described.

The most widely accepted system for classifying air masses uses the thermal characteristics of the source regions: tropical (T), polar (P), and less frequently Arctic or Antarctic (A). The moisture characteristics of the air mass are represented by the words continental (c) and maritime (m) corresponding to dry and humid air, respectively. In this classification an air mass formed over a tropical ocean is called maritime tropical and is labeled mT; a continental polar air mass, cP, is cold and dry and originates over a continental area at high latitudes.

When an air mass leaves its source region, it gradually changes as a result of interactions with the underlying surface and vertical air motions. Air which is warmer than the surface over which it is passing is identified by adding the letter w. Thus maritime tropical air moving over a cold continent is designated as mTw. When the air is colder than the underlying surface, it is identified with the letter k. A polar continental air mass moving southward over warmer land is identified as cPk.

As noted in Chapter 5, the vertical stability of air depends mostly on how the temperature varies with height. A k-type air mass tends to be unstable because the warm surface produces a steep lapse rate and convection. As a result, the air generally is well mixed by vertical motions and therefore visibility through the air tends to be good. The daylight sky has a deep blue color when viewed through fresh, clean cPk air.

On the other hand, w-type air masses generally are stable near the ground because the air is warmer than the cold surface. The resulting low-level temperature inversion suppresses vertical mixing. In such circumstances, air pollutants are trapped and vertical mixing is restricted. Suspended particles scatter sunlight in such a fashion as to give the sky a whitish cast.

The major source regions of various air masses affecting North America are shown in Fig. 7-3. As noted earlier, they are large areas having generally uniform surface conditions.

Continental polar air masses originate over the snow- and ice-covered land masses of North America and Eurasia. The chief processes in the formation of cP and cA air are radiation and condensation. The white surfaces reflect much of the incoming solar rays and at the same time are efficient emitters of heat in the form of infrared waves. The air close to the ground radiates heat both upward to the sky and downward to the snow surface which, in turn, radiates skyward. Some of the energy is absorbed by the air, but most of it is lost to outer space. As a result, the temperature of the air close to the ground decreases. The process continues as a deeper and deeper layer of cold, stable air is produced. When the temperature of the surface

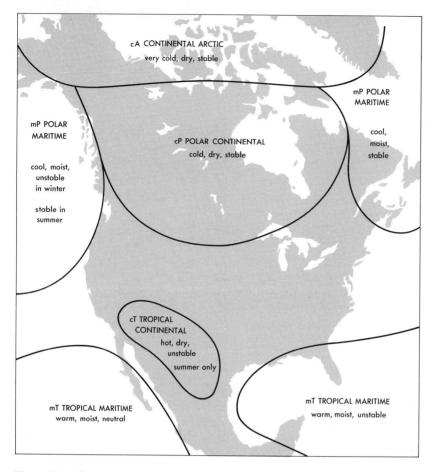

Figure 7-3 Air mass source regions over and around North America.

falls low enough, water vapor condenses on the ground. This reduces the water vapor content of the air and leads to the characteristic low humidity in cP and cA air.

In some cases, the cooling process can go on for several weeks and produce a huge region of cold, dry air 2000 m to 3000 m deep. The vertical distributions of temperature (T) and dew point (T_D) in a continental polar air mass are shown in Fig. 7-4. Note that in a qualitative sense the smaller the differences between T and T_D, the higher the relative humidity. This sounding was made in the early morning hours before sunrise and shows an extremely stable temperature inversion in the lowest 200 m. Two lines displaying the dry adiabatic lapse rate have been included to allow ready judgments of the stability of the various layers of the atmosphere. It is evident that above 2.5 km the lapse rate was substantially less than the adiabatic and therefore was stable. The layer from 1.0 km to 2.5 km was only slightly stable.

128

The water vapor content of the air was very low (no more than about 1.3 g/kg of air). This amounts to about 1.4 g/m³ of air It will be seen that even over desert areas water vapor contents in the air are far greater than the amounts in this cP air. The chief reason for the low moisture values is found in the low temperatures. Note that the relative humidity at the ground was a fairly high 62 percent, but because the temperature was −8.2°C, the air could hold only about 2.3 g/m³ even if the relative humidity were 100 percent.

In winter most of the northern part of the massive land mass of Eurasia is a source of continental polar air. Over the Asian northland it can be bitterly cold. For example, Verkhoyansk in Siberia has winter temperatures as low as −68°C (−90°F). This frigid air, flowing outward over Europe and Asia behind the polar front, can dominate winter weather in the Eastern Hemisphere. Continental arctic air masses are an extreme form of continental polar air. They form over polar ice and are very cold and dry.

When a cP or cA air mass moves over the ocean, it can be converted to a maritime polar (mP) air mass in only a day or two. Regions downwind

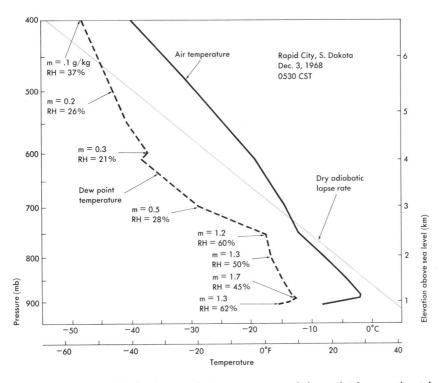

Figure 7-4 The vertical distribution of air temperature and dew point in a continental polar (cP) air mass. Values of humidity at various points are shown in terms of the mixing ratio, *m*, the number of grams of water vapor in a kilogram of air.

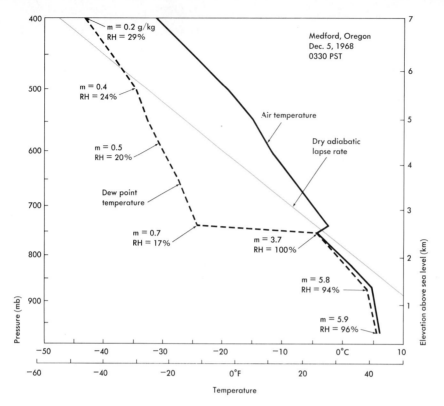

Figure 7-5 The vertical distribution of temperature and humidity in a maritime polar (mP) air mass.

of oceans such as the British Isles and western Europe commonly are under the influence of mP air. The underlying water, having a higher temperature than the air, warms it by conduction. This leads to instability and convection currents that rapidly transport heat upward. The convection also transports water vapor evaporated from the sea surface into the dry air. Figure 7-5 shows the structure of a maritime polar air mass. It was very humid through the lowest 2.5 km, the layer through which convective mixing transported water vapor from the ocean surface. Above the temperature inversion at 2.5 km the air was dry and stable.

Most of the summer showers and thunderstorms over the United States develop in maritime tropical (mT) air from the Gulf of Mexico, the Caribbean Sea, and the nearby Atlantic Ocean. In the summer the Indian Ocean, India, and southeast Asia are also dominated by mT air. The ocean waters in these regions have relatively high temperatures and as a result they warm the air moving over them. At the same time, water vapor is evaporated in large quantities. Convection mixes heat and water vapor vertically through the lowest layers of the atmosphere. Figure 7-6 shows that the mT air sounded in this case was warm, humid, and unstable up to the base of the inversion caused by subsiding air.

130

In the winter as maritime tropical air moves northward over colder land, it becomes mTw air. Since temperature increases with height, the air is stable near the ground, but aloft the air remains unstable and contains a great deal of water vapor. When such air is lifted, as it might be over a frontal surface, the released latent heat can lead to upward buoyancy. The resulting clouds can yield large quantities of rain or snow.

The maritime tropical air that forms over the Pacific Ocean just west of Mexico is warm and humid, but it is not as unstable as its counterpart over the Gulf of Mexico. The ocean waters off the west coast of Mexico are relatively cool and as a result there is a tendency toward low-level stability.

Figure 7-3 shows that continental tropical (cT) air forms over the southwestern United States and Mexico. Much larger source regions in the Northern Hemisphere are the desert areas of northern Africa and southern Asia. Continental tropical air masses are mostly a summer phenomenon and are

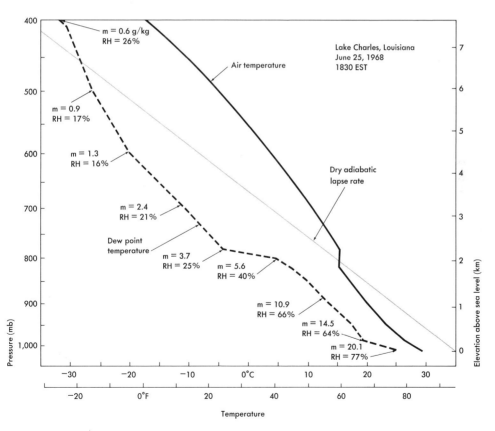

Figure 7-6 The vertical distribution of temperature and humidity in a maritime tropical (mT) air mass.

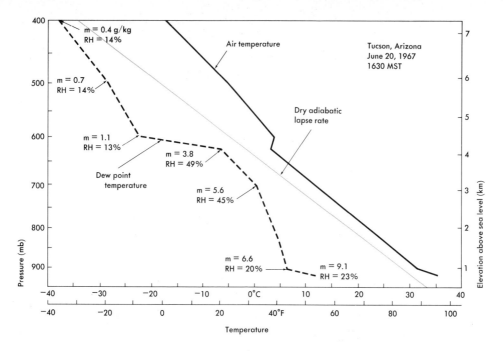

Figure 7-7 The vertical distribution of temperature and humidity in a continental tropical (cT) air mass.

hot, dry, and unstable. Insolation causes air temperatures in the deserts to reach extremes that may exceed 40°C; soil temperatures may be 10°C to 20°C higher. In the lowest few hundred meters of the atmosphere the temperature lapse rate can be greater than the dry adiabatic lapse rate. The air in the lowest layers of the atmosphere is very unstable and there is a great deal of clear air convection, often up to altitudes exceeding 3000 m. These features are illustrated in Fig. 7-7 which displays temperature and humidity over Tucson, Arizona, on a summer afternoon. There was a superadiabatic layer through the first 200 m above the ground and a deep, nearly adiabatic layer extending to the base of a temperature inversion at 4 km. Through this region the air was well mixed by convective overturning. The temperature inversion was a result of subsidence of air on the eastern edge of the large semipermanent anticyclone that normally extends eastward from over the Pacific Ocean. Above the inversion, because of subsidence and compression effects, the air was much drier than it was just below the stable layer. It can be seen that the relative humidity decreased from 49 percent at the base of the inversion to 13 percent through a vertical distance of about 300 m.

Near the ground the actual quantity of water vapor was about 9.1 g in 1 kg of air. This amounts to about 9.2 g/m³, a fairly high value. Nevertheless, because of the high air temperature of about 36°C, the relative humidity was only 23 percent.

132

Pilots who have flown over the deserts during summer afternoons are well acquainted with the convection currents in cT air. They serve to mix pollutants released at the ground through a deep layer. At night over the deserts there usually is strong outgoing infrared radiation. As noted in Chapter 5, this leads to the development of shallow, ground-based temperature inversions that trap atmospheric contaminants. Fortunately for desert dwellers, the incoming solar rays "burn off" the inversion and generally bring about a deep mixing layer every afternoon. This sequence of events acts to increase the desert atmosphere's capacity for pollutants, but it does not make it unlimited. Even with a mixing layer 3 km deep, there can be a serious degradation of atmospheric quality if the quantity of pollutants is sufficiently high.

FRONTS

When air masses having different properties come together, they do not mix readily. Instead, the warmer, less dense air overrides the cooler, more dense air. As shown in Fig. 7-8 there is a zone of transition between the air masses. It is called a front, a term brought into meteorology during World War I when opposing armies faced one another across a battle front. Along horizontal surfaces through the front the temperature gradients are large compared to the temperature gradients within either air mass.

As a result of observations of atmospheric conditions at the surface and aloft, it has been possible to identify various types of fronts. Every one of

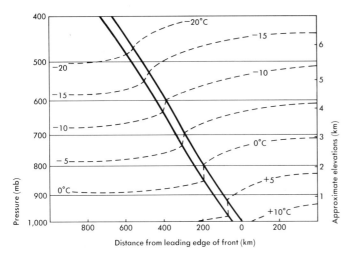

Figure 7-8   Vertical cross section through lower part of a front showing an enlarged view of the air temperature pattern.

them separates cold air from warmer air. When the pressure distribution is such as to cause the cold air to advance and the warm air to retreat, the zone of transition is called a *cold front* (Fig. 7-9). On weather charts fronts are drawn as curved lines, but it should be recognized that they represent sloping zones of transition whose horizontal widths might be 10 km to 100 km. Typically, the wedge of cold air has a slope of perhaps 1 km of rise for 100 km of distance. When a cold front advances rapidly, warm air can be forced to rise just ahead of the front. If the warm air is unstable, convective clouds, perhaps even thunderstorms, can occur along the leading edge of the frontal zone.

Figure 7-9(B) shows the pressure and wind patterns in the vicinity of a typical cold front. Winds in the warm air blow from the southeast nearly

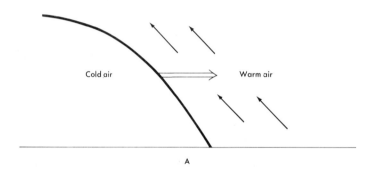

A

Figure 7-9 A cold front: (A) a vertical cross section with an exaggerated vertical scale; (B) example of winds and pressure on a surface map. The double arrows show the direction of motion of the front.

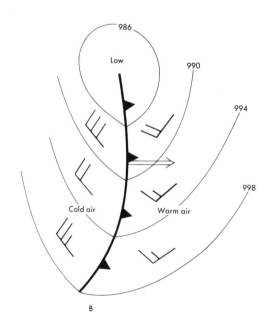

B

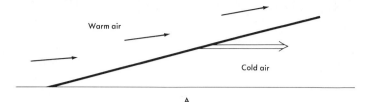

Warm air

Cold air

A

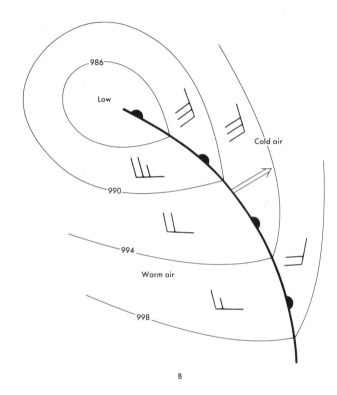

986

Low

Cold air

990

994

Warm air

998

B

Figure 7-10 A warm front: (A) a vertical cross section with an exaggerated vertical scale; (B) example of winds and pressure on a surface map. The double arrows show the direction of motion of the front.

parallel to the isobars. When the front passes, there is a sudden wind shift. Often the winds increase in intensity, blow from the northwest, and are gusty. At the same time, temperatures fall as the cold air replaces the warm air.

When the wind directions are such as to cause the warm air to advance while the cold air retreats, the zone of transition between them is called a *warm front* (Fig. 7-10). Usually, the slope of a warm front is substantially less than that of a cold front. The warm air tends to glide up the front and produce widespread cloud systems. Sometimes the transition zone between warm and cold air does not move one way or the other. In such cases it is called a *stationary front*.

On occasion a cold front overtakes a warm front and leads to the formation of an *occluded front*. Figure 7-11 illustrates a cold-type occlusion

135

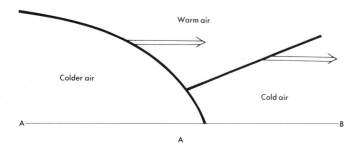

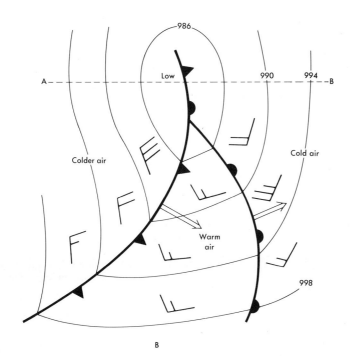

Figure 7-11 An occluded front. The vertical cross section in drawing (A) is along the line AB in drawing (B).

because the air under the cold front is colder than that under the warm front. When the reverse is true, the cold front overrides the warm front and produces a warm-type occluded front. As the occlusion process continues, the warm air is forced to rise and extensive cloud systems with rain or snow are the result.

A front of particular importance is the one that separates tropical air from polar air. As shown in Fig. 7-12, this zone of transition, called the *polar front*, sometimes extends in a wavy pattern almost continuously around the North Hemisphere. At other times the front exists as broken segments around the globe.

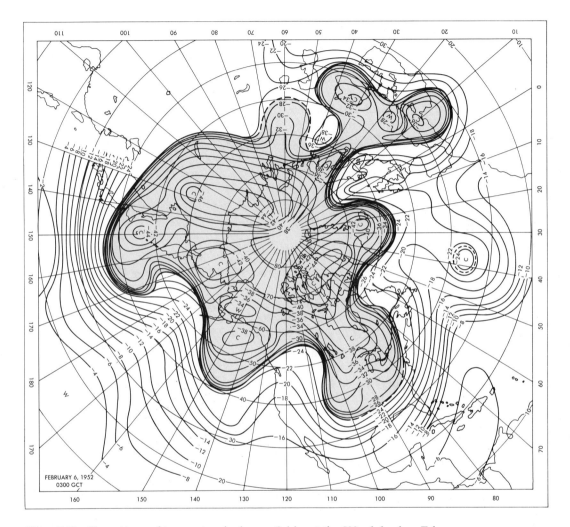

Figure 7-12 The patterns of temperature in degrees Celsius at the 500-mb level on February 6, 1952, over the Northern Hemisphere. The polar front, shown by the heavy line, coincides with the southern edge of the zone where the isotherms are close together. The shaded area represents the region of cold polar air. Note that in the middle Atlantic there is an isolated region of cold air. It would be said to be "cutoff" from the principal polar air mass. *From* D. C. Bradbury and E. Palmén, *Bulletin of American Meteorological Society*, 1953, **34**: 56–62.

CYCLONES

The frontal zone is a particularly interesting region because of the marked changes of air properties across it. This point is illustrated in Fig. 7-8 which is a schematic representation of temperature and pressure on each side of a cold frontal zone. Note that on each side, in the relatively uniform air masses, the levels of constant pressure and the levels of constant temperature are essentially parallel to each other. Since density depends mostly on temperature, surfaces of constant density would also tend to be parallel to surfaces of constant pressure. When this is the case, the atmosphere is said to be *barotropic*. When isobaric and constant density surfaces are not parallel, as is the case in the frontal zone, the atmosphere is said to be *baroclinic*. These two conditions are a measure of the overall stability of the atmosphere as far as the formation of cyclones is concerned. This type of stability is sometimes referred to as *hydrodynamic stability*. Before discussing this point, let us examine the interactions of fronts and cyclones as revealed by weather map analyses.

As long ago as the early part of the nineteenth century it was recognized in England that large storm systems were associated with passing regions of low pressure, commonly called *depressions* in the European literature. For about a hundred years various scientists speculated about the properties of such cyclonic storms. Unfortunately, there was a paucity of reliable observations taken over a sufficiently large region and an absence of systematic measurements in the upper atmosphere. This made it difficult to construct a reasonably complete picture of what was happening in storm centers.

Major advances in the understanding of cyclones were made toward the end of World War I by the Norwegian meteorologists Vilhelm Bjerknes, his son Jakob, and associates. They collected sufficient synoptic observations to study the structure of a number of cyclones over Europe. The result of their analyses is the now famous *frontal theory of cyclones*. It was widely accepted for about three decades as an adequate explanation of cyclone development. Over recent decades more complete theories for cyclone formation have been formulated.

According to the Norwegian school, a cyclone forms along a nearly stationary front following a pattern often observed on weather maps. The sequence of events is shown in Fig. 7-13. A wave appears on the front as cold air is deflected southward and warm air is deflected northward. In the process, a cyclonic circula ion is initiated and the kinetic energy of the system increases as cold, heavy air behind the front sinks while warm, light air rises.

As the wave develops, condensation occurs in the warm, ascending air, clouds form, and precipitation may occur. At the apex of the wave there is a fall in pressure that produces a wave cyclone containing a warm front and cold front. The cold front advances faster than the warm front, and the sector of warm air becomes progressively smaller. Recall that air motion around a Northern Hemisphere cyclone is counterclockwise.

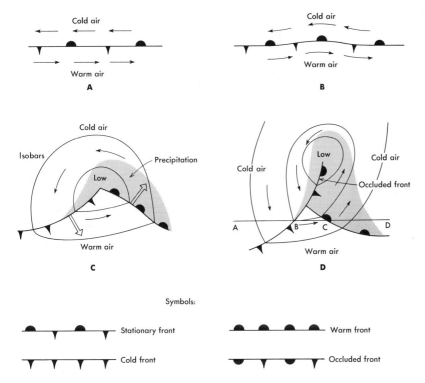

Figure 7-13 Schematic representation of the formation of a wave cyclone along a frontal zone. The pattern of clouds along the cross section *ABCD* in (D) is shown in Fig. 8-10 of the next chapter.

As the cyclone continues to evolve, the advancing cold front overtakes the warm front and occlusion occurs as shown in Fig. 7-13. Under the occluded front there is a mixture of the cold air that was below both the cold and warm fronts. Continued occlusion leads to dissipation, and in the final stage the cyclone is a weakening vortex comprised of fairly uniform air.

As a Northern Hemisphere cyclone advances from the west and passes over a particular place, there is a predictable sequence of weather events. The pressure falls as the low-pressure system approaches. At the same time the wind gradually turns from east toward the south. Such a clockwise turning of the wind is called *veering*. Under the warm front the skies generally are overcast and may be yielding rain or snow. The passage of the warm front is accompanied by an increase in temperature, a veering of the wind into the westerly quadrant and a clearing of clouds. Skies in the warm sector often have a milky blue appearance as a result of light scattering by particles suspended in the air.

As the cold front approaches there may be convective clouds and the pressure will be falling. The cold frontal passage usually will be accompanied by gusty winds increasing in speed and turning into the northwest. Temperatures will fall, and skies are likely to clear and become bluer because of the relatively clean nature of cold, polar air.

Figure 7-14 Observation of the north Pacific Ocean at 1945 GMT on September 10, 1976, showing a well-developed cyclone centered at 47°N latitude and 135°W longitude off the west coast of the United States. The east–west zone of clouds at about 10°N latitude is associated with the intertropical convergence zone. *Courtesy* National Oceanic and Atmospheric Administration.

The clouds produced by cyclones can be seen best from weather satellites. As shown in Fig. 7-14, sometimes the cloud pattern bears a close resemblance to the one pictured in Fig. 7-13, with a spiral pattern of clouds around the low-pressure system and a long trail of clouds along the cold front.

In reality, the sequence of cyclone development may differ in many important details from the idealized pattern illustrated in Fig. 7-13. In particular, the pattern of precipitation is often very different from those depicted. Most of the weather is found ahead of the center of low pressure and is associated with the warm front, but there are many variations from one storm to another.

The life cycle of a cyclone may consume from a day to a week depending on the degree of development. The first stages depicted in Fig. 7-13(A) and

7-13(B) may occur in hours. Typically, along a long stationary front many small wave disturbances occur, but only a small number of them develop into mature cyclonic storms. When one does occur, another is likely to occur along the same stationary front. This leads to what is known as a *cyclone family*.

The frontal theory of cyclones has been a very useful concept, and it is still widely used by weather forecasters for predicting the formation and behavior of cyclonic weather systems.

The frontal theory, however, does not lead to a satisfactory mathematical model for the formation of cyclones on a sloping frontal surface.

In the late 1940's a number of scientists, particularly Jule Charney, now at the Massachusetts Institute of Technology, began investigating cyclone development starting from a different point of view. This research has led to a concept known as the *baroclinic wave theory of cyclones*. As noted earlier, the word "baroclinic" is used to specify a state of the atmosphere where surfaces of constant pressure are not parallel to surfaces of constant density. Since density depends mostly on the temperature, in a baroclinic atmosphere the temperature varies along a constant pressure surface (Fig. 7-15). Note from Fig. 7-8 that a frontal region is baroclinic. Typically, in the troposphere the temperature decreases from south to north. In such a case, the north–south pressure gradient and the west–east wind velocity increase with altitude. When the wind increases markedly with altitude and when other conditions are satisfied, a condition known as *baroclinic instability* develops, and cyclonic storms may form. In such a circumstance, certain pressure and wind perturbations, once started, continue to grow in amplitude.

The baroclinic wave theory has a solid mathematical foundation in explaining the early stages of cyclone formation. Research in the early 1950's indicated that in typical circumstances when clouds and precipitation did not occur, the most unstable north–south wind perturbations were those occurring in wave-like patterns having wavelengths of about 4000 km. They could double their amplitude, that is, the north–south excursions of the wind pattern, in about 40 hours.

The condensation of water vapor and the release of latent heat in rising air have an important effect on the preferred wavelength and the time of

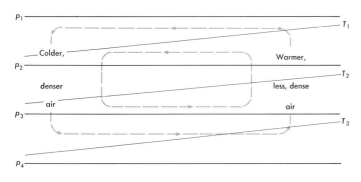

Figure 7-15 When the temperature and pressure surfaces are not parallel, the atmosphere is baroclinic and there is a tendency for the circulation shown by the arrows.

development. If the initial perturbation produces a region of clouds 5 km by 150 km, wave cyclones having wavelengths of about 1000 km would be the ones most likely to occur. Furthermore, storm amplitudes would double in about 14 hours. The sizes and growth rates calculated from theory are in reasonable agreement with observations.

According to the baroclinic theory of cyclone development, the north–south perturbations of wind velocity are accompanied by vertical velocities as relatively warm air ascends and the colder air descends (Fig. 7-15). Where air is rising, there must be an inward flow of air to feed the updrafts. As the air converges, it develops a cyclonic spin at increasing speeds because of the effects of the earth's rotation. According to the principle of the conservation of angular momentum, as the mass of a system gets closer to the axis of rotation, the rotational velocity increases. This is analogous to the actions of an ice skater who brings her arms closer to her body in order to spin faster.

In areas of sinking air the air diverges and its spin weakens; sometimes an anticyclonic circulation develops. As noted earlier, in the Northern Hemisphere this means a clockwise flow around a high-pressure center.

As an incipient cyclone intensifies, the ascending current strengthens and causes condensation and latent heat release. This, in turn, leads to still greater intensification and continued pressure falls in the cyclone. The consequence is a region of closed, nearly circular isobars that are characteristic of cyclones.

The baroclinic theory explains the formation of cyclones without reference to the existence of a frontal boundary between cold and warm air. One of the attractive features of the baroclinic theory is that it treats cyclones and anticyclones as an essential part of the general circulation. They form in the westerly flow in middle latitudes because other north–south exchange mechanisms are inadequate to transport energy northward. As the north–south temperature gradient increases, the middle-latitude flow becomes unstable. It breaks down into cyclonic and anticyclonic waves that serve to transport heat poleward.

It should be noted that the baroclinic wave theory was developed through the use of mathematical techniques dealing with small perturbations in initial conditions. With the advent of high-speed computers and the development of numerical analysis methods, it has become possible to use theoretical models to describe the entire history of a cyclone. With some exceptions, the approach is similar to the mathematical treatment of the general circulation mentioned in Chapter 6. For example, for the study of cyclone development over a period of several days, it is not necessary to take into account atmosphere–ocean interactions and the conditions over the entire earth. The set of equations specifies changes with time of the state of the atmosphere. It the initial state is known, it is possible to calculate future ones in a step-wise fashion from one time to the next. Such a procedure is called *numerical weather prediction*. More will be said about it in Chapter 12.

DISTRIBUTION
OF CYCLONES
AROUND THE WORLD

The analysis of large numbers of weather maps shows that certain regions are more favored for cyclone development than others. In the Southern Hemisphere most cyclones occur between 50°S and 60°S latitudes. There is relatively little difference in frequency of occurrence around the globe or from summer to winter. On the other hand, in the Northern Hemisphere there are large seasonal differences in frequency and preferred locations. In summer there is a pronounced maximum in cyclone frequency at about 60°N latitude.

As would be expected, during the winter the polar fronts move southward, and cyclones are found most often at about 50°N. Many cyclones form along the frontal boundary that extends from southeast Asia to Alaska, spanning most of the North Pacific Ocean (see Fig. 7-16). As the depressions develop and move northeastward, they sometimes become unusually intense and come to rest in the Gulf of Alaska. This accounts for the pronounced low-pressure system known as the *Aleutian low* which appears on climatological charts of sea-level pressure (Fig. 6-2).

Many cyclones also form over the mid-Pacific and take a southerly course toward the west coast of the United States. As they pass over the high mountain barriers along the western rim of North America, the cyclones often lose their characteristic pressure and wind structures and are difficult to follow, but many of them redevelop on the east side of the mountains. One prominent region where this happens is to the east of the Sierra Nevadas. A particularly favored location for cyclone regeneration is over eastern Colorado. Such storms sometimes are called *Colorado lows*. As shown in Fig. 7-17, they tend to move toward the Great Lakes and increase in intensity as a result of the interaction of polar air and tropical air from the south. These storms are the source of heavy, widespread rain and snow. Another region where cyclone redevelopment is often observed is just to the east of the Canadian Rockies. These storms are sometimes called *Alberta lows*, after the province over which they occur. As they intensity and move eastward or southeastward, they can bring blizzards with blowing snow and bitter cold air.

Cyclones are commonly observed in a broad band extending across the north Atlantic from the Gulf of Mexico to Europe. The band coincides with the average position of the polar front in this part of the world. A favored cyclone spawning region is along the south and east coast of the United States. Some cyclones form in the Gulf of Mexico and move northward on the west side of the Appalachians. More often they move over the Gulf Stream toward Iceland where they tend to stagnate. This accounts for the *Icelandic low* seen on climatological charts of sea-level pressure (Fig. 6-2).

Many cyclones develop along a zone extending from Iceland to the Barents Sea. These storms move eastward over northern Europe and sometimes as far as Siberia.

As is the case over the Great Lakes, heat given off by large regions of warmer water in the winter serves to stimulate cyclone formation. This explains the high frequency of cyclone occurrence over the Baltic, Black, and Caspian seas. Large numbers of depressions form over the northern

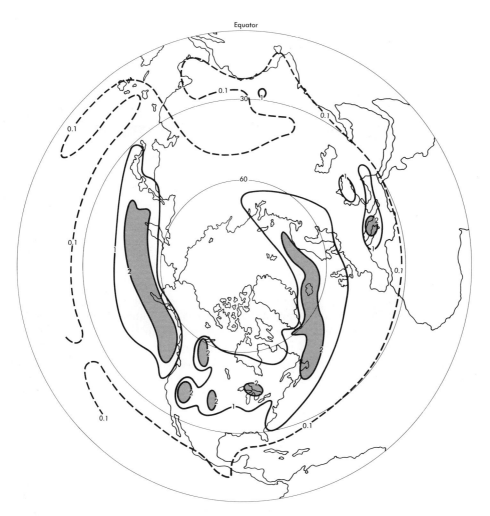

Figure 7-16 Relative frequency of cyclone centers in winter. The numbers refer to the percentage frequency of occurrence within areas of 10^5 km² obtained from the fraction of days with a cyclone present to the total numbers of days counted. Weak cyclones lasting less than 24 hours were not included in the analysis. *From* S. Petterssen, *Introduction to Meteorology*, 3rd ed., McGraw-Hill, 1969.

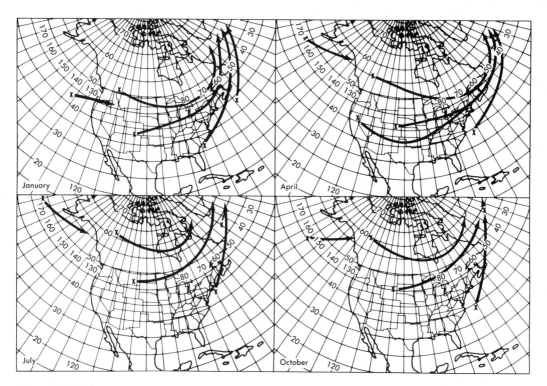

Figure 7-17 Primary cyclone tracks over North America during different seasons of the year. Each of the *X*'s identifies a region where cyclone formation is relatively frequent. *From* C. H. Reitan, *Monthly Weather Review*, 1974, **102**: 861–868.

edge of the Mediterranean in the winter. They usually move northeastward or eastward, sometimes as far as northern India, and give winter rainfall.

In the summer cyclones associated with the baroclinicity of frontal zones are less frequent than in the winter. They most often occur over the north Pacific and north Atlantic Oceans. One observes high frequencies of low pressure centers over Arizona, California, northern Mexico, northern Africa, and Arabia. These cyclones are known as *thermal lows* or *heat lows*. They exist over these desert areas because the very high air temperatures in the lowest layers of the atmosphere result in low overall air densities and low pressures. At upper levels there is usually an anticyclone and subsiding air. Heat lows are mostly stationary and may persist for many weeks; they are not associated with fronts and the cyclonic wind circulation is generally weak. Figure 7-7 shows a temperature and moisture sounding through continental tropical air in a heat low. In this case, the low humidities and subsidence aloft inhibit the formation of clouds and rain. When the air in a heat low is humid, showers and thunderstorms are common, especially in mountainous areas.

In the summer over India there is a high frequency of cyclones associated with the monsoon. Since the air is hot, humid, and unstable, widespread heavy rain is a common feature of these storms.

Over the warm oceans at lower latitudes in late summer there are many cyclones that generally travel from the east while in the tropics. Sometimes they reach great intensities, become hurricanes and exhibit strong wind and rain, and produce ocean waves of huge proportions. Their characteristics will be discussed in Chapter 9.

CHAPTER 8

Clouds, Rain, and Snow

FOR CENTURIES clouds have had special meanings for many people. Poets, painters, and composers have found inspiration in the beauty of their forms and colors. Farmers have looked up to them gratefully when they produced rain to feed thirsty crops. Other farmers have shaken their fists at the clouds when they sent down hail or strong winds that destroyed grains or fruits. Meteorologists look on clouds with less emotion, but what they see is no less interesting.

CLOUD TYPES

Even a casual observer is aware of the great variety of clouds. They range in size from small, white puffs to dark, gray layers extending from horizon to horizon. Some last for minutes; others persist for hours; some storm systems go on for days. On some occasions clouds move rapidly over the landscape; at other times, particularly over mountainous terrain, they hardly move at all, seemingly held by an invisible anchor to a nearby ridge. Certain clouds are uniform in texture, density, and brightness while others exhibit wavy patterns and many shadings from white to gray.

The nature of a cloud depends on a number of factors, the most important being the pattern of vertical motions and the moisture content and temperature of the air. When the air is unstable and updrafts are strong and persistent, clouds of great vertical development are the rule. If the air rises slowly and steadily over a large region, a layer-type cloud occurs. Often there are wave motions that propagate horizontally on stable layers in the atmosphere. Although invisible, they produce patterns of vertical motion resembling the waves on the surface of a lake when a stone is dropped into it. The clouds are indicative of the existence of a wave-like pattern of updrafts.

Over the years a number of procedures have been devised for classifying clouds. The most widely used classification was introduced by an English chemist, Luke Howard, in 1803 and is the one adopted by the World Meteorological Organization (WMO). For the most part, the clouds are classed according to their appearance. In the WMO system there are ten cloud *genera* based on the main characteristic forms of clouds (Table 8-1). Each of the

Table 8-1 CLOUD GENERA ACCORDING TO WORLD
METEOROLOGICAL ORGANIZATION

Cloud genera	Description
Cirrus ⎫ Cirrocumulus ⎬ Cirrostratus ⎭	Mostly ice crystal, low temperatures, high altitudes (3 km to 18 km, depending on latitude)
Altocumulus ⎫ Altostratus ⎬ Nimbostratus ⎭	Water or ice clouds, middle altitudes (2 km to 8 km, depending on latitude)
Stratocumulus ⎱ Stratus ⎰	Water or ice clouds, low altitudes (below 2 km)
Cumulus ⎱ Cumulonimbus ⎰	Water or ice clouds, vertical development

genera comes in one or more of 14 *species* depending on peculiarities in the shapes or internal structures of the clouds. In addition, they are further subdivided according to the arrangement of their parts, transparencies, and other distinctive features. There is little point in this book to go into a detailed discussion of the great variety of clouds, but it is in order to learn some general features of the most common clouds. An examination of Table 8-1 suggests that the cloud genera can be subdivided into three main classes —cumulus, stratus, and cirrus.

Cumulus are detached clouds having the appearance of rising mounds, domes, or towers (Fig. 8-1). They are formed by convective currents whose structure can be visualized by the white, cauliflower appearance of the upper parts of the clouds. As a cumulus cloud enlarges, it is given the name *cumulus congestus*. Finally, when the cloud becomes exceptionally dense and vertically developed, with a dark base from which rain may begin to fall, it becomes a *cumulonimbus* (Fig. 8-2). Such a cloud may have its upper parts in the shape of an anvil. It often produces lightning and thunder and therefore is some- times called a *thundercloud* or a *thunderstorm*.

Figure 8-1 Cumulus clouds.

A

Figure 8-2 The development of a cumulonimbus cloud with anvil. The photographs were taken at the following times: (A) 1356 MST (B) 1401 MST (C) 1406 MST (D) 1411 MST (E) 1416 MST.

B

C

D

E

Figure 8-3 Altostratus cloud. *Courtesy* National Center for Atmospheric Research, Boulder, Colorado.

As the name implies, clouds in the stratus family are arranged in flat layers. When the cloud appears as a gray, uniform sheet at an elevation below a kilometer or two, it is called *stratus*. If a cloud having this appearance occurs at higher altitudes, say 2 km to 7 km over middle latitudes, it is called *altostratus* (Fig. 8-3). When rain or snow is falling from a stratified cloud, the word *nimbus* is used as a prefix to obtain the name *nimbostratus*. Incidentally, when the water or ice particles evaporate before reaching the ground, the precipitation is called *virga*.

When many cumuliform cloud elements are arranged in a layer and the cloud is below about 2 km, the composite is called a *stratocumulus* (Fig. 8-4). If the cloud layer is between about 2 km and 7 km in temperate regions of the earth, the individual convective elements appear relatively small, and the cloud is called *altocumulus* (Fig. 8-5). Over mountain areas you sometimes see a beautiful, almost stationary cloud having the shape of a giant lens or even a stack of lenses one on top of the other (Fig. 8-6). Such a cloud, called *altocumulus lenticularis*, is formed as a result of wave motion on the lee side of a ridge. The air flow is as shown in Fig. 8-7. Along the upwind side cloud droplets are steadily forming while at the downwind edge they are steadily evaporating. As a result, the cloud position stays essentially fixed

Figure 8-4 Stratocumulus formed by the spreading out of cumulus. *Courtesy* National Oceanic and Atmospheric Administration.

151

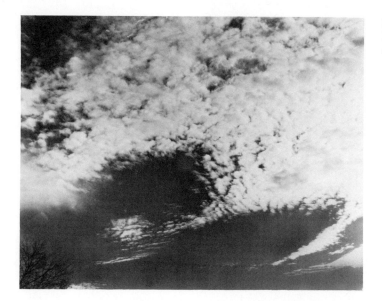

Figure 8-5 Altocumulus. *Courtesy* National Oceanic and Atmospheric Administration.

Figure 8-6 Altocumulus clouds shaped like a giant lens and therefore called *altocumulus lenticularis*. *Courtesy* National Oceanic and Atmospheric Administration.

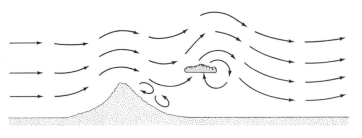

Figure 8-7 Mountain waves are formed when strong winds blow across a mountain ridge. Strong updrafts and downdrafts, accompanied by severe turbulence, occur downwind of the mountain. Lenticularis clouds such as those in Fig. 8-6 may form in the rising air.

with respect to the mountain barrier. When the airflow is strong and the air has the appropriate stability, the vertical motions just downwind of the ridge can be very strong and extend to altitudes exceeding 10 km. The air can be extremely turbulent and can be a hazard to aviation.

Cirrus clouds such as those shown in Fig. 8-8 are made up mostly of ice crystals and they usually occur at high altitudes—5 km to 13 km in middle latitudes. These clouds are often in the form of white, delicate filaments. When a cirroform cloud appears as a whitish veil that is fibrous or smooth and covers much or all of the sky, it is called *cirrostratus*. Such a cloud sometimes produces a *halo*, a circular ring or arc of light that circles with the sun or the moon (see Fig. 10-7).

A high layer of clouds made up of many very small elements having the appearance of grains or ripples is called a *cirrocumulus* (Fig. 8-9). Clouds

Figure 8-8 Cirrus clouds.

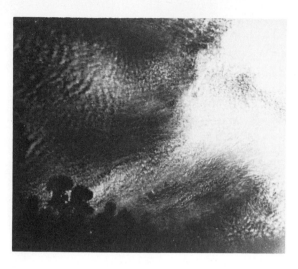

Figure 8-9 Cirrocumulus clouds. *Courtesy* National Oceanic and Atmospheric Administration.

of this type sometimes take on spectacular reddish colorations as the sun sets. The blue component of sunlight is scattered out of the rays as they pass through long distances of cloudless air. As a result, cirrocumulus (or small-element altocumulus) clouds may resemble the scales of a mackerel, and the phrase *mackerel sky* is used to describe the view.

Although fog is not usually considered to be a cloud, it has essentially the same properties. The two phenomena differ only to the extent that the base of fog is at the ground while clouds are above the ground. According to international definition, a fog is a suspension of minute water or ice particles that reduces visibility, at the surface, to distances less than 1 km. In a mountainous region a layer of cloud observed by a person in a valley would be a fog to someone on a ridge engulfed in a sea of water droplets.

Occasionally, the sequence of clouds passing overhead is a good indication of the weather changes likely to occur during the next day or two. Long before government weather services came in to being, sailors and farmers used wind and cloud observations to predict the weather. Such observations are particularly informative during the approach of well-developed cyclones in middle latitudes. The reasons for this are illustrated in Fig. 8-10 which shows a vertical cross section through a cyclone such as the one depicted in Fig. 7-13(D).

As the warm front approaches, the surface winds are generally from the southerly quadrant; cirrus clouds appear high in the sky. They are slowly replaced with a layer of cirrostratus. It thickens, and lowers, and becomes altostratus as the surface warm front gets closer. Perhaps a day after the first sighting of the cirrus, rain or snow starts falling from the lowering overcast of thick nimbostratus clouds.

With the passage of a surface warm front, the winds shift abruptly to the southwest and temperatures rise. The sky clears gradually. Air in the

154

warm sector may be fairly humid. Sometimes scattered cumuliform clouds and cumulonimbus occur. The appearance of patches of altocumulus at high elevations signals the approach of a cold front. Organized bands of showers and thunderstorms often occur ahead of or along the surface cold front. When it passes, the wind shifts to the northwest, generally increases, and becomes gusty. Temperatures may fall rapidly and skies clear as cold, dry air passes over the point of observation.

If every cyclonic storm followed this simple pattern, weather forecasting would be easy. In fact, patterns of winds, clouds, and precipitation vary from storm to storm. In some cases, cyclones move rapidly and the sequence of weather events may take a day or two. In other cases, storms may stagnate over a particular region, and rain or snow may go on for nearly a week.

Before moving on to an examination of the properties of cloud droplets, mention should be made of the interesting clouds occurring in and above the stratosphere. The existence of clouds at great altitudes is somewhat surprising because of the low concentrations of water vapor. Upper atmosphere clouds not only have been scientific puzzles over the years, but they also have been a source of great wonder and admiration because of their striking beauty. *Nacreous clouds*, made up of tiny ice particles, occur at 20 km to 30 km and are mainly observed at northern latitudes in the winter (Fig. 8-11). They are commonly called *mother-of-pearl clouds* because of their brilliant pearly colorations when viewed during and just after sunset. As the sun dips below the horizon and dark sets in, clouds in the lower atmosphere also darken because they are in the shadow of the earth. On the other hand, the high level clouds can still be brightly illuminated by the rays of the sun. In the morning hours nacreous clouds are sometimes observed just before sunrise.

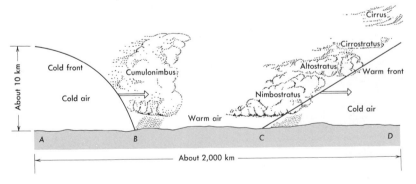

Figure 8-10 Clouds associated with a fully developed cyclone. This vertical cross section is taken along the line *ABCD* in Fig. 7-13(D). Note that the horizontal and vertical scales are very different.

Figure 8-11 Nacreous clouds. The bright lens-shaped clouds exhibit beautiful mother-of-pearl colors. The thin veil of clouds is uniform in color. *Photo by* Y. Gotaas; *courtesy* E. Hesstvedt, University of Oslo, Norway.

At the great altitudes of 75 km to 90 km where temperatures can be below −120°C *noctilucent clouds* are observed on rare occasions (Fig. 8-12). They are thin clouds thought to be made up of ice crystals that have grown on dust particles. They have been sighted most often during the summer months in both hemispheres in the latitude belts 50°N to 75°N and 40°S to 60°S. Noctilucent clouds are seen at twilight when the sun is below the horizon and its rays strike the cloud particles and are reflected toward the earth. Just after sunset the clouds have a grayish color, but they become more brilliantly bluish-white as the night progresses. In the summer, at the latitudes involved, twilight comprises most of the summer night. As sunrise approaches, the clouds become gray again.

Figure 8-12 Noctilucent clouds at Watson Lake, Yukon Territory, Canada. *Courtesy* B. Fogle, National Science Foundation.

PROPERTIES OF CLOUD DROPLETS

A cloud is a collection of minute water or ice particles sufficiently numerous to be seen. Most cloud droplets have diameters ranging from a few to about 100 μm, but occasionally some will be perhaps twice as large. Figure 8-13 is a photograph of cloud droplets captured on a microscope slide exposed to the air as an airplane flew through a cloud. Most of the droplets were about 10 μm to 20 μm in diameter. As will be seen in a later section, most clouds contain enormous numbers of droplets.

The cloud droplets appearing in the photograph are almost perfect spheres. They are so small that surface tension forces acting toward the centers of the droplets overcome other forces and lead to spherical shapes. Later in this chapter it will be shown that as the water drops in clouds grow larger and become raindrops, their shapes depart substantially from sphericity.

It might be asked what the difference is between a cloud drop and a raindrop. The primary distinction is one of size. The larger the size, the faster the speed of fall of a drop and the greater the distance it will fall before it evaporates. These points are shown in Table 8-2. Drops of diameters smaller than 200 μm fall very slowly, and when they descend out of a cloud they evaporate very quickly. On the other hand, drops of diameters exceeding 2000 μm (2 mm) fall relatively fast and descend several kilometers below the cloud base before evaporating. As a consequence, they reach the ground in the form of raindrops. Meteorologists have decided that for most purposes it is convenient to use a diameter of 200 μm as the demarcation between cloud

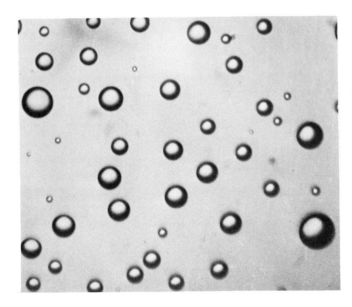

Figure 8-13 Photograph of cloud droplets captured on an oil-coated microscope slide while flying through a cloud. The largest droplets are about 40 μm in diameter.

Table 8-2 FALL VELOCITY OF WATER DROPS IN STILL AIR AT SEA LEVEL

	Diameter (μm)	Fall velocity (cm/s)	Fall distance for complete evaporation (m)*
Cloud droplets	1	0.003	
	5	0.076	
	10	0.30	
	20	1.0	Less than 1
	50	7.6	
	100	27	
Drizzle drops	200	72	150
	500	206	
	1000	403	
Raindrops	2000	649	4200
	3000	806	
	5000	909	

*Data was calculated for the following conditions: pressure = 900 mb, temperature = 5°C, and relative humidity = 90%. Note that 100 cm/s = 3.28 ft/s = 2.24 mi/hr.

Source: R. J. List, *Smithsonian Meteorological Tables*, 6th ed., 1958.

droplets and raindrops. The table shows that water drops having diameters between 200 μm and 500 μm are called *drizzle drops*. Although this is the correct name for this size range, it is often ignored and drops of diameters greater than 200 μm are regarded as raindrops.

It might be surprising to some readers to learn that the fall speed of spheres increases with their diameter. This is further illustrated by Fig. 8-14 which depicts the quantity called the *terminal velocity* plotted as a function of water sphere diameter. Imagine the sequence of events occurring when a sphere is dropped. It begins to fall under the influence of the force of gravity and accelerates at a rate of 9.8 m/s^2, the value of gravitational acceleration. If there were no other forces, after 1 sec of fall the droplet would be falling at a speed of 9.8 m/s (22 mi/hr). In fact, as soon as the droplet begins to move, it encounters air molecules that exert a frictional force. It is usually called a *drag force* and it is directed upward and increases as the speed increases. At a certain point in the droplet's fall the upward drag force becomes equal to the downward gravitational force. When this happens, the acceleration equals zero. As a consequence, the droplet's fall speed attains a constant value which is the terminal velocity.

Calculations of the net forces on falling water spheres show that the terminal velocities vary with diameter in the manner shown in Table 8-2 and Fig. 8-14. As will be seen later, the fact that the large drops fall faster than the small ones is important in explaining the formation of rain. At this time let us consider the properties of cloud droplets and how they grow.

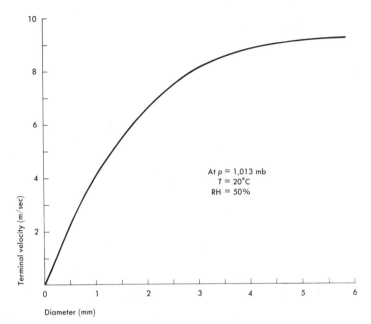

At p = 1,013 mb
T = 20°C
RH = 50%

Figure 8-14 Terminal velocity of raindrops at about sea level. *From* R. Gunn and G. D. Kinzer, *Journal of Meteorology*, 1949, **6**: 243–248.

The spectrum of cloud-droplet sizes varies from one cloud type to the next and among clouds of the same species. As a matter of fact, from one place and time to others in the same cloud, there often are large differences in the droplet characteristics. But even with all these variations, it can be shown that, on the average, the droplet properties within any particular cloud type are somewhat distinctive. Figure 8-15 shows how the size spectra of droplets differ in three types of clouds in the cumulus family.

Cumuli of fair weather are small, white, puffy clouds. They seldom are more than a kilometer in vertical extent; often they are less than 300 m thick. They never produce rain. It can be seen that in this kind of cloud the droplets are small and numerous. Their diameters usually are smaller than 40 μm and the concentrations of droplets are greater than 300 per cubic centimeter.

In large cumulus clouds that produce showers over the central United States it is found that many droplet diameters exceed 50 μm. The concentrations of droplets are smaller than those in fair-weather cumuli; they may be about 200 per cubic centimeter.

The cumulus clouds over tropical oceans have droplet characteristics differing markedly from those found in continental clouds. Unlike their overland counterparts, tropical cumuli produce rain when they are relatively small. Observations over the Caribbean Sea show that the cloud bases usually are at an altitude of about 600 m, and when the cloud tops extend to 3.0 km, rain is often seen to fall from them. Cumulus clouds over continents seldom produce rain unless their summits rise above about 5 km. The cloud droplets

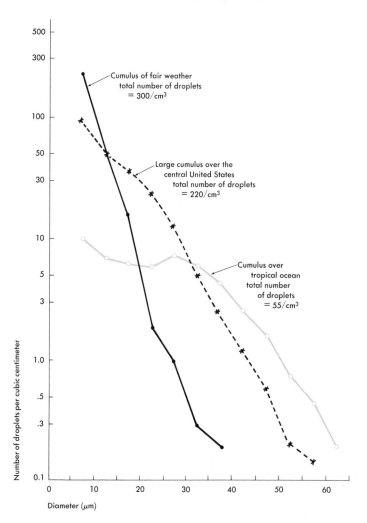

Figure 8-15 Examples of measured concentrations of droplets as a function of diameter in three types of cumuliform clouds. Each point on each curve represents the number of droplets having diameters within a 5-μm interval.

in tropical clouds occur in concentrations of only about 60 per cubic centimeter, but there are numerous droplets having diameters greater than 60 μm. It has been suggested that the presence of many large cloud droplets in shallow clouds can be attributed to a large number of so-called *giant* particles of salt on which the droplets grow. Such salt particles can have diameters of several micrometers.

In stratiform clouds the droplets are usually smaller than in cumuliform clouds. Average droplet diameters range from about 10 μm to 20 μm while the concentrations may be from about 200 to 600 per cubic centimeter. In shower-producing cumulus clouds the average diameters are from 20 μm to 40 μm, and concentrations are 50 to 200 per cubic centimeter.

THE GROWTH
OF CLOUD DROPLETS

The formation of cloud droplets as a result of condensation is a familiar, but not always recognized, sight. On cold days you can clearly see the process at work by merely breathing out and watching clouds of tiny water droplets appear as the water vapor you exhale condenses on particles in the air. There are many other examples, such as the steam cloud above the chimney of a steam locomotive and the pearly white trails produced by high-flying jet airplanes.

Condensation is a process whereby water vapor molecules are caused to come together in sufficient numbers to produce liquid water. When the molecules condense on large surfaces, such as blades of grass, we need only consider the nature and temperature of the surface and the temperature and relative humidity of the air in order to understand how condensation comes about.

Consider a glass of ice water on a day when the relative humidity of the air is 50 percent and the temperature is 30°C. When the glass of water is first set on a table, the air in contact with the glass begins to cool because the temperature of the glass is 0°C, that of a water-and-ice mixture. For reasons discussed in Chapter 5, as the air cools, its relative humidity increases.

After a period of time, depending mostly on the initial moisture properties of the air, the air in contact with the glass attains a relative humidity of 100 percent and is saturated. Further cooling causes the air to become supersaturated. Water vapor molecules begin to deposit themselves on the surface of the glass in order to restore the air to the saturated state. As long as saturation is exceeded, there will be a continuation of the condensation process. In the case of a glass of ice water, the air in the vicinity of the glass is replaced as air moves around it. Consequently, the condensation process goes on and on, and the glass continues dripping water on the table. If there were a limited supply of air, the process would slow down as the temperature of the air surrounding the glass and the supply of moisture in the air decreased.

As was noted in Chapter 5, the temperature of the air at which condensation just begins is the *dew-point temperature*. It depends on the air temperature, moisture content, and pressure. In the case already cited (when the temperature is 30°C and the relative humidity is 50 percent) the dew-point temperature at sea-level pressure is 18°C.

On a fairly clean metal or glass surface condensation begins when its temperature equals the dew-point temperature of the air. In the case of minute particles, especially if they are composed of substances having an affinity for water vapor, the situation is more complicated.

When tiny droplets, such as those having radii less than about 10 μm are involved, the shape—specifically the curvature—of the droplet affects the saturation vapor pressure over it. In order for such small droplets to be in equilibrium with the surrounding air, it is necessary for the relative humidity to be greater than 100 percent. For example, if only pure water were involved, it would be necessary to have relative humidities of over 140 percent in order to prevent a droplet of diameter 0.006 μm from evaporating immediately. When droplet diameters exceed about 10 μm, the equilibrium relative humidity approaches 100 percent.

In the atmosphere relative humidities seldom exceed 101 percent, and even these 1 percent supersaturations occur only in strong updrafts in thunderstorms. How then can cloud droplets form? The answer is that they grow on particular types of minute particles which are called *condensation nuclei*. Typically, they have diameters of about 0.1 μm to 3 μm or more and are found in concentrations of a few 100 per cubic centimeter of air. Condensation nuclei having diameters greater than about 2 μm are called *giant nuclei*. As shown in Tables 2-2 and 8-3, the atmosphere contains huge numbers of particles, but most of them are too small to be condensation nuclei at the humidities occurring in the atmosphere.

Condensation nuclei come from many sources, particularly blowing soil, volcanoes, smokestacks, and the oceans. In addition, nuclei are formed in the atmosphere as a result of chemical reactions involving gases such as sulfur dioxide and nitrogen dioxide. The most favorable nuclei are *hygroscopic*, that is, they have a marked ability to accelerate the condensation of water. Examples of hygroscopic nuclei are acid particles and sea salt. Condensation on ordinary table salt, sodium chloride, may begin when the relative humidity is only 75 percent. Magnesium chloride is even more hygroscopic, and condensation can start with relative humidities below 70 percent.

The role played by a hygroscopic salt nulceus in the growth of a cloud droplet is shown in Fig. 8-16. The dashed curve represents the relative humidities, measured with respect to a plane water surface, that are needed to maintain in equilibrium pure water droplets having the radii shown on the horizontal scale. It can be seen that as the droplet's size decreases, higher supersaturations are needed in order to prevent it from evaporation.

Table 8-3   Size Distribution of Particles
                    in the Atmosphere

Nuclei	Diameter (μm)	Concentration (particles/cm³)
Aitken	Less than 0.2	10^4 to 10^5
Large	0.2–2	10^2
Giant	Greater than 2	1

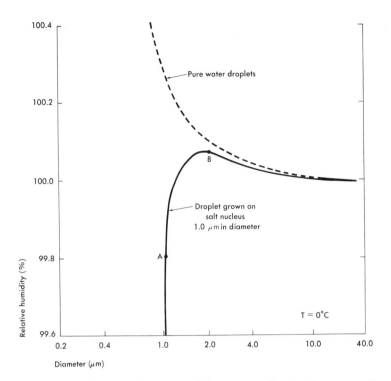

Figure 8-16 The equilibrium relative humidity (that is, the relative humidity at which a droplet, whose diameter is given by the horizontal scale, remains the same size) depends on the size and the composition of the droplet. If the actual relative humidity is greater than the value shown on the curve, the droplet grows by condensation. If the relative humidity is less than that given by the curve, the droplet evaporates.

The solid curve shows the relative humidities needed to maintain equilibrium over a droplet growing on a salt nucleus of 1.0 μm in diameter. When such a nucleus exists in air having a relative humidity of 99.8 percent, condensation occurs and a droplet grows to the size indicated by the dot labeled A. At this point, because of the water condensed on it, the droplet will be in equilibrium with its surroundings and will remain this size unless the relative humidity is changed. If it were increased to 100.1 percent, there would be a resumption of condensation and droplet growth. Once the droplet size "passes over the hump" (point B) on the curve, it continues to grow as long as the relative humidity exceeds 100 percent. By the time the droplet has grown to a size greater than about 10 μ, the curvature and salt-solution effects are so small that the droplet acts almost in the same manner as would a large surface of pure water. When this happens, the situation is similar in some respects to that of the glass of ice water. Condensation continues as long as the air is supersaturated with respect to the pure water.

The process of condensation can be readily explained in terms of vapor pressures. The vapor pressure is a measure of the pressure exerted by water molecules. If the temperature is constant, the greater the number of water molecules in a given volume of air, the higher the vapor pressure. One can also speak of the vapor pressure at the surface of a liquid; it is equal to the saturation vapor pressure of the air just at the surface of the liquid and depends only on temperature. When the air over a water droplet is supersaturated, that is, when its relative humidity is greater than 100 percent, the vapor pressure of the air is higher than the saturation vapor pressure at the drop surface. As a result, there is a pressure force driving the water molecules from high to lower pressure, namely, from the air to the water. This is the condensation process at work.

In nature, relative humidities are increased in various ways. For example, evaporation of raindrops falling into a dry layer of air can sometimes saturate it and cause clouds to form. Air moving over a body of water is humidified by evaporation. At the same time, if this air is warmer than the water, it can lose heat to the water and have its relative humidity increased as the air temperature falls. Such a process can lead to fogs over the sea.

Land fogs are commonly found on clear nights when moist air near the ground loses heat to outer space as a result of radiation. Such fogs usually are first observed in low places because the cooler, heavier air drains downward in much the same fashion as water running downhill.

Most clouds, particularly those producing rain or snow, occur as a result of ascending currents of air that are cooled adiabatically in the manner described in Chapter 5. Within a rising body of air there is usually a large population of particles that can serve as condensation nuclei. As the air cools and the relative humidity increases, the likelihood of condensation increases. The height of the base of convective clouds indicates the amount of vertical motion and cooling required to lead to saturation and condensation.

The largest hygroscopic nuclei are activated first and grow as water molecules condense on them. As water accumulates the nuclei dissolve and the equilibrium vapor pressure over these drops approaches that of pure water. As a consequence, smaller hygroscopic nuclei become effective and begin to grow. This process goes on and leads to cloud droplet spectra of the type shown in Fig. 8-15. The numbers and sizes of droplets depend on the sizes and compositions of the original condensation nuclei and the characteristics of the updrafts within which the droplets grow.

The collisions of large and small cloud droplets and consequent coalescence cause further growth of the larger droplets. This causes a broadening of the size spectra and accounts in part for the differences in the spectra shown in Fig. 8-15. As will be seen later in this chapter, a continuation of the collision and coalescence processes can lead to rain.

ICE CLOUDS

Although cloud droplets form on condensation nuclei, the water making up a full-grown droplet is surprisingly pure because there is very much more water than nuclei material. For example, if a 50-μm drop formed on a 0.1-μm nucleus, the volume of water would be about 125,000,000 times greater than the volume of the nucleus.

The purity of water in clouds is important because it accounts for the fact that many clouds are made up of water droplets even when their temperatures are below the nominal freezing point at 0°C. Such clouds are called *supercooled*. As will be seen later, supercooled clouds serve as the physical basis for many techniques of cloud and weather modification that employ cloud seeding.

Not all clouds are made up of water droplets. Most clouds found at altitudes where temperatures are below 0°C contain ice crystals. Even clouds that are initially supercooled generally are converted to ice by natural causes when they reach sufficiently low temperatures.

Extremely pure water can be supercooled to about −40°C. At lower temperatures ice forms without the presence of foreign substances in the water. At higher temperatures, especially at about −5°C to −20°C where ice crystals usually occur in the atmosphere, the ice formation is started by particles called *ice nuclei*. Table 8-4 lists the temperatures at which minute particles of various substances produce ice crystals. The first one listed, solid

Table 8-4 THRESHOLD TEMPERATURES
AT WHICH VARIOUS SUBSTANCES
PRODUCE ICE CRYSTALS

Substance	*Temperature* (°C)
Dry Ice (solid CO_2)	0
Silver iodide	−4
Lead iodide	−6
Naturally occurring:	
Covellite	−5
Vaterite	−7
Magnetite	−8
Kaolinite	−9
Illite	−9
Haematite	−10
Dolomite	−14
Montmorillonite	−16

Source: B. J. Mason, *The Physics of Clouds,* Oxford University Press, 1972.

carbon dioxide, or Dry Ice as it is often called, does not really serve as an ice nucleus. When pellets of Dry Ice are dropped into a supercooled cloud, they produce ice crystals by lowering the temperature of the air. Solid carbon dioxide has a temperature of about $-78°C$. When the humid air in a super-cooled cloud is cooled below about $-40°C$, ice crystals are formed without a foreign substance serving as nuclei. When ice crystals form this way, the process is called *homogeneous nucleation.*

Natural ice nuclei are thought to come chiefly from the ground. Certain clays such as kaolinite and montmorillonite are abundant and reasonably effective nuclei because they initiate the ice process at temperatures between $-9°C$ and $-16°C$. As shown in Table 8-4, certain substances produce ice crystals at warmer temperatures, but they are scarce in nature. As will be seen in Chapter 12, Dry Ice and silver iodide are the materials most frequently introduced into supercooled clouds in the course of weather modification experiments. Once ice crystals have formed in a cloud, the number can be increased by the splintering of existing crystals.

Crystals can take on a great variety of shapes depending mostly on the temperature and vapor pressure of the air (Fig. 8-17). A notable feature of

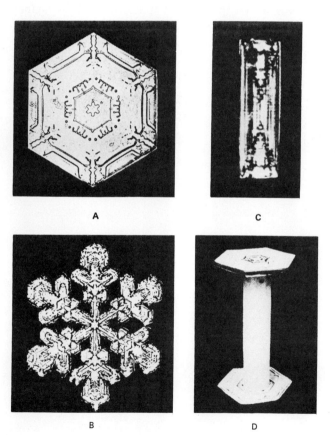

Figure 8-17 Various types of ice crystals found in the atmosphere. (A) plate; (B) dendrite; (C) column; (D) column capped with plates. *From* W. A. Bentley and W. J. Humphreys, *Snow Crystals*, Dover Publications, Inc., 1962.

A

C

B

D

ice crystals is their hexagonal construction that is attributed to the hexagonal structure of the water molecule. Some crystals are long hexagonal columns; others are flat hexagonal plates. The most beautiful ones are the dendrites that come in an almost infinite variety of patterns with intricate structures that are a pleasure to behold.

FORMATION OF RAIN AND SNOW

Although condensation readily produces cloud droplets, in normal circumstances it cannot account for the formation of raindrops. Once a cloud droplet reaches a diameter of a few tens of microns, the water is so pure that condensation can continue only if the air is supersaturated with water vapor. But as the air approaches saturation, there are numerous small, less-favored condensation nuclei that can be activated. In such a circumstance, the available water vapor is condensed on the smaller, newly formed droplets instead of on the existing larger ones. As long as there are large numbers of condensation nuclei available, and this is almost always the case, the available water vapor condenses on so many particles that few exceed diameters of about 50 μm and most of them are much smaller. Raindrops usually are 10 to 100 times larger. Figure 8-18 illustrates the size of raindrops in relation to a condensation nucleus and cloud droplets. The diagram also gives representative particle concentrations in numbers per cubic meter and fall velocities in still air.

For many purposes, you can assume typical cloud droplet and raindrop diameters to be 20 μm and 2000 μm (2 mm) respectively and typical concentrations of cloud and raindrops to be 100 droplets/cm³ and 100 drops/m³, respectively. Since the volume of a sphere is $(\pi/6)$ (D^3), where D is the diameter, the quantity of water in such a single raindrop is equivalent to the water in about a million such cloud droplets.

There are two principal mechanisms by which precipitation particles are produced in nature. The first one, called the *coalescence process*, involves the collision and merging of water droplets falling at different velocities. This occurs because the large cloud droplets fall faster than the small ones (Table 8-2) and sometimes collide with them. The probability of collision depends on the relative sizes of the droplets. When there are some as large as 40 μm to 50 μm in diameter in a cloud made up mostly of droplets smaller than 20 μm, the frequency of collision is significant and increases as the larger drops grow.

Interestingly, laboratory experiments show that in some circumstances water droplets collide without coalescing. Instead, they bounce off one another. Apparently a very thin layer of air molecules prevents the water molecules from coming into contact, and surface tension forces give the droplets the elasticity allowing large deformations without breaking. Some experi-

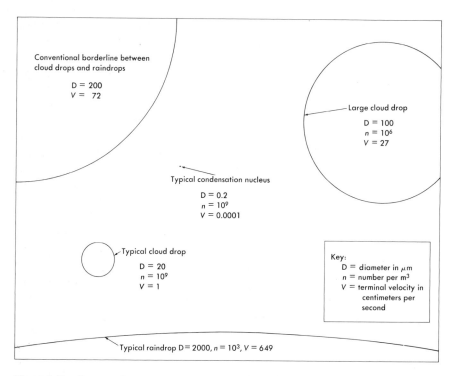

Figure 8-18 Comparative diameters, concentrations, and terminal fall velocities of some particles involved in condensation and precipitation processes. Note particularly the great difference in diameter of a typical cloud drop and of a typical raindrop. *From* J. E. McDonald, *Advances in Geophysics*, Academic Press, 1958.

ments have shown, however, that relatively small electrical forces can cause essentially all collisions to be followed by coalescence. The forces are present when the droplets are carrying electric charges or when they find themselves in an electrified cloud such as a developing cumuliform cloud (see Chapter 9).

There is no doubt that in relatively small convective clouds over the tropical oceans, rain is produced entirely by the coalescence process. Clouds having their bases at an altitude of about 600 m are often observed to rain when their tops reach 3 km where the temperature is about +7°C. The coalescence process also plays a part in the production of rain in many other clouds, but its importance is difficult to evaluate because some other process may also be at work.

Much of the earth's precipitation falls in the form of snow or raindrops that come from melting snowflakes. An examination of snowflakes shows them to be large ice crystals or, more commonly, aggregates of ice crystals. The process by which they form was studied in some detail about 1930 by the famous Scandinavian meteorologist Tor Bergeron and later extended by

W. Findeisen in Germany and others as well. Precipitation formation by means of the *ice-crystal process* is based on the fact that if ice crystals and supercooled water droplets coexist at subfreezing temperatures, the crystals grow while the droplets evaporate. This occurs because, as shown in Fig. 8-19, the saturation vapor pressure over water is greater than the saturation vapor pressure over ice at the same subfreezing temperature. As a result, there is a pressure force driving water molecules from water to ice.

As noted earlier, supercooled clouds often occur in the atmosphere. If ice crystals are introduced into such clouds, perhaps by the action of ice nuclei of the type listed in Table 8-4, the stability of the cloud system is suddenly changed. The ice crystals grow rapidly as the water droplets evaporate. In some cases, the crystals reach diameters of a few hundred micrometers in a few minutes. As they grow, they begin falling through the cloud and colliding with supercooled droplets and other ice crystals. The droplets may freeze on contact; the crystals stick to one another. In this way, snowflakes may be produced.

If air temperatures are low enough, the snowflakes reach the ground. On many occasions, however, the air near the ground has temperatures high enough above freezing to allow the snowflakes to melt and rain occurs. In mountainous areas it is common to see snow at the high elevations and rain in the valleys.

The range of sizes of raindrops depends on the processes by which they are produced and the properties of the cloud in which they developed. In deep cumulonimbus clouds having strong and widespread updrafts ice and water particles can grow to fairly large sizes. As raindrops enlarge, they take on shapes very different from spherical. This point is illustrated in the

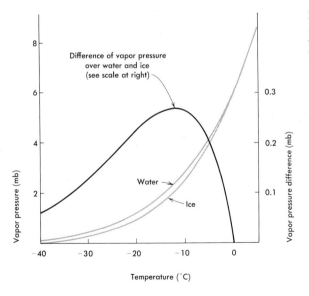

Figure 8-19 Saturation vapor pressure over supercooled water and ice at the same subfreezing temperature (scale on left) and the difference between them (scale on right).

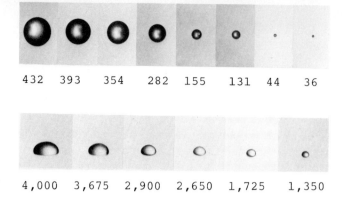

| 432 | 393 | 354 | 282 | 155 | 131 | 44 | 36 |

| 4,000 | 3,675 | 2,900 | 2,650 | 1,725 | 1,350 |

Figure 8-20 Photograph of water drops in air showing changes of drop shape with increasing size. The diameters in micrometers are indicated under each photograph. Note that 1000 μm = 1 mm. *From* H. R. Pruppacher and K. V. Beard, *Quarterly Journal of the Royal Meteorological Society*, 1970, 96: 247–256.

series of photographs in Fig. 8-20. Note that large drops exhibit a flat bottom and a rounded top. They more nearly resemble hamburger rolls than the teardrop shapes often pictured by artists. As large drops fall through turbulent air and collide with other drops, they oscillate and change their shapes. Sometimes the deformations become so large that the drops rupture and form a number of smaller ones. The bigger the drop, the greater the chances that it will break up. This puts an upper limit of about 5 mm on the diameter of raindrops. The largest drops fall from thunderstorms, often at the start of the rain. It is suspected that they result from the melting of large ice particles.

There are a variety of instruments used to measure raindrop sizes and frequency. Some are optical devices that view the drops as they fall; others sense the momentum of the falling drops as they impinge on a suitable probe; still others observe the size of spots made by raindrops falling on and spreading across a treated surface. Incidentally, as a rough rule of thumb, when a raindrop hits and spreads across smooth cement, such as a sidewalk, the spot it leaves has a diameter about seven times larger than the diameter of the original raindrop.

In some winter storms strong temperature inversions occur in the lowest kilometer or two of the atmosphere and may lead to a particularly hazardous form of rain (Fig. 8-21). The snowflakes melt as they fall through the warm layer having a temperature greater than 0°C. The resulting water drops then become supercooled as they pass through the cold air near the ground. On striking the cold pavement and other objects such as automobiles, trees, and power lines, the supercooled water freezes rapidly. This *freezing rain* coats everything with layers of hard, solid ice called *glaze*. Heavy accumulations of ice may cause widespread damage to vegetation; slick roadways present very dangerous driving conditions (Fig. 8-22).

When partially melted snowflakes or supercooled raindrops freeze and become solid grains of ice, the precipitation is called *sleet*. Sometimes the news media use the same term to represent a mixture of snow and rain. This meaning of sleet is the accepted one in British terminology and is used colloquially in various parts of the United States. Sleet particles are part of a class of transparent or translucent ice particles called *ice pellets*. They have diameters 5 mm or smaller; when diameters exceed 5 mm, the precipitation is called *hail*.

Figure 8-21 (Right) Snowflakes melt as they fall through the warm layer where temperatures exceed 0°C. As liquid raindrops descend through the layer of subfreezing air near the ground, they may become supercooled and freeze on impact with the cold ground.

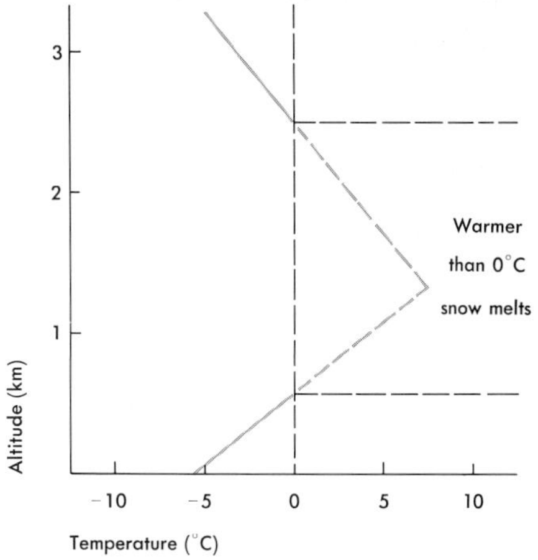

Figure 8-22 (Left) The effects of freezing rain on power lines. *Courtesy* National Oceanic and Atmospheric Administration.

HAIL

Hail has been observed at one time or another in almost every place where thunderstorms occur. But in some areas it is very seldom seen, while in others hail is common. The average number of thunderstorms and hailstorms each year over the United States is shown in Fig. 8-23. Florida experiences more thunderstorms than any other state, but the occasions of hail at the ground are few. Over the western Great Plains of the United States the frequency of thunderstorms is less than over Florida, but the number of hailstorms is much greater.

Other regions of the world also have high frequencies of damaging hailstorms. Northern Italy and the Caucasus region of the U.S.S.R., South Africa, Kenya, and Argentina are among them.

Sizes of hailstones vary over a wide range, with stones over 2 cm in diameter being not unusual in regions of high hail frequency. In extreme cases,

171

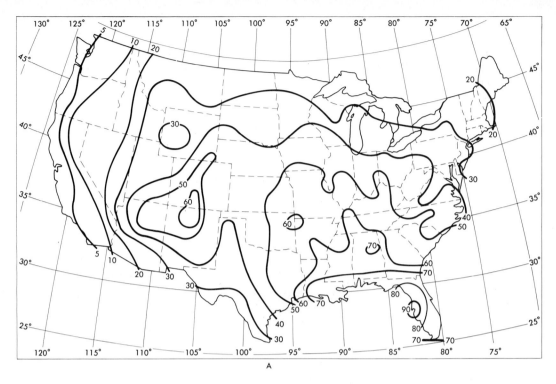

A

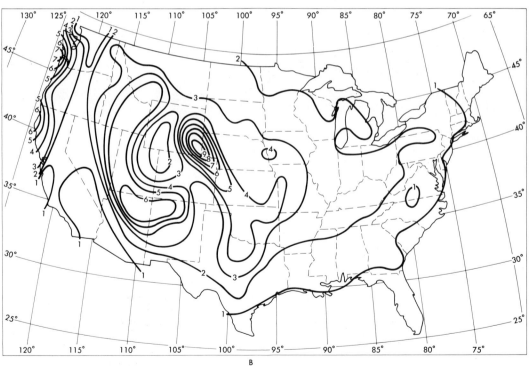

B

Figure 8-23 (Left page) The geographical distributions of thunderstorms (A) and hail-storms (B) in the United States. The numbers represent the average number of days per year at any point over a 40-year period. From *Thunderstorm Rainfall*, Hydrometeorological Report No. 5, Department of Commerce.

hailstones may exceed the diameter of an orange. In September, 1970, a hailstone about 14 cm in diameter and weighing 766 g (1.67 lb) fell on Coffee-ville, Kansas (Fig. 8-24). As seen in this photograph, hailstones can have shapes that are far from being spherical. They have been found to be conical, ellipsoidal, or amorphous. It is easy to imagine how giant hailstones pose a threat to vegetation and property, but most damage to agriculture, such as that illustrated in Fig. 8-25, is done by heavy falls of relatively small hail, particularly when the winds are strong.

A common property of hailstones is that the ice of which they are made up is not uniform in texture. Some parts of almost every stone are made up of clear ice while other parts are milky ice. The opaque appearance is caused by trapped air bubbles. In large stones alternate layers of clear and opaque ice are found sometimes.

Details of the interior structure of hailstones have been studied by slicing thin sections and examining them under various types of light. Figure 8-26 is a photograph of a thin section of hailstone viewed under polarized light. The stone is made up of many thousands of ice crystals. Each crystal has a flat face that makes a particular angle with the rays of light passing through the thin slice of ice. A large crystal appears as a large area of uniform shade of gray; a small crystal appears as a small area of uniform shade. The

Figure 8-24 The record size hailstone which fell at Coffeeville, Kansas, on September 3, 1970. It had a circumference of 44 cm and a mass of 766 g (1.67 lb). *Courtesy* National Center for Atmospheric Research, Boulder, Colorado.

Figure 8-25 The effects of a hailstorm on a corn field. *Courtesy* United States Department of Agriculture.

region of opaque ice is made up of small crystals and trapped air bubbles, while the region of clear ice is made up of large crystals. The center of the hailstone shown in Fig. 8-25 was made up of many large crystals (clear ice). It was surrounded by a thin layer of small crystals (opaque ice) and then a thick layer of large crystals (clear ice).

Why should there be alternate layers of large and small crystals? The explanation is found in the rate at which the water is collected and frozen on the growing hailstone. When it falls through a region of the cloud containing small concentrations of supercooled water, the hailstone intercepts small quantities of water that may freeze almost instantly. In the process, tiny air bubbles are trapped in the ice and small crystals are the rule. On the other hand, if the falling stone accumulates large quantities of water, it cannot freeze instantly. Instead, the stone gets wet, and the freezing proceeds slowly as large crystals grow. In the process, the air is forced out of the water.

In attempting to reconstruct the series of events leading to the formation of hailstones, it is necessary to account for their sizes as well as the layers of clear and opaque ice. It is known that thunderstorms that produce hail are accompanied by tall cumulonimbus clouds containing strong updrafts. The quantity of supercooled cloud water must be high in order to allow for rapid growth of the stones. This quantity usually has magnitudes between 0.01 g/m^3 and 1.0 g/m^3 cloud air, but in thunderstorms it has been observed to be as high as 4 g/m^3 to 5 g/m^3. In theory, it is possible to obtain much higher values.

Figure 8-26 The crystal structure of a hailstone revealed when a thin slice is viewed under polarized light. The core of this hailstone was made up of clear ice in the form of large crystals, surrounded by a shell of milky ice made up of small crystals. A thicker shell of clear ice formed the outer layer. *Courtesy* National Center for Atmospheric Research, Boulder, Colorado.

Hail may fall from various types of thunderstorms, from relatively small isolated ones or storms that cover large regions and last for many hours. Research over the last decade or so shows that certain thunderstorms having tilted updrafts may be prolific hail producers.

A highly regarded theory of hail formation, first offered by two English scientists, Frank H. Ludlam and Keith A. Browning, visualizes the following sequence of events (Fig. 8-27). Hailstones begin to form in the rising air entering the cloud along the leading edge and containing supercooled cloud droplets in its upper reaches. As the stones grow, they begin to fall at increasing speed through the air and intercept increasing quantities of supercooled water. If the updraft speed is high and exceeds the terminal velocity of the hailstones, they are carried upward. The upper level winds move them forward across the upper part of the storm. When the updraft is not strong enough to hold up the hailstones, they fall toward the ground. After descending several kilometers, some of the stones fall into the updraft current and experience a second passage through the supercooled updraft. A series of such trips can lead to large hailstones. An ice particle starting with a diameter of 1 mm can grow to 3 cm after exposure to the high liquid water contents in a strong updraft. Clear ice forms as the hailstone accumulates large quantities of supercooled water. As the hailstone falls through cold air outside the updraft core, it encounters low quantities of liquid water. As a result, the ice develops an opaque appearance. More will be said about hailstorms in Chapters 9 and 12.

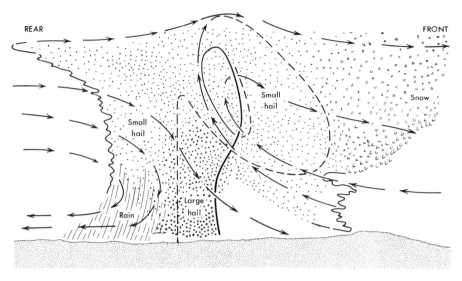

Figure 8-27 Simplified version of the physical model of severe hailstorm proposed by Keith A. Browning and Frank A. Ludlam. Arrows show air motions with respect to the cloud.

HYDROLOGIC CYCLE

Since an adequate supply of water is essential if life as we know it is to prosper, earth scientists have given a great deal of attention to the exchanges and transformation of the water medium between atmosphere, oceans, and continents. The composite picture is called the *hydrologic cycle* (Fig. 8-28).

Although there is a great mass of water on the earth, most of it is heavily loaded with salts and is contained in the oceans. This is shown in Table 8-5. As far as human beings are concerned, the most vital water supplies are the relatively small quantities contained in rivers, lakes, ground water, and the water that circulates between the atmosphere and the continents.

In this chapter we have dealt with a part of the hydrologic cycle—the formation of clouds and precipitation. In Chapter 3, another part of the picture, namely, evaporation from the oceans, was examined. Quantitative data on the depths of water evaporated from and precipitated on the oceans and continents are given in Table 8-6. The total runoff is the difference between precipitation and evaporation. A complete treatment of the hydrologic cycle could easily occupy an entire chapter, but space permits only a brief outline of certain important points.

Over the entire earth the average yearly quantity of water added to the atmosphere as a result of evaporation amounts to a depth of about 100 cm.

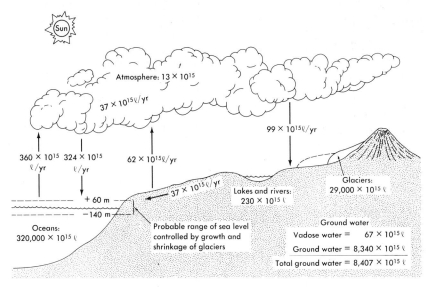

Figure 8-28 The hydrosphere and the hydrologic cycle. One liter (ℓ) equals 10^3 cm^3 and 0.26 gal. *From* B. J. Skinner, *Earth Resources* 2nd ed., 1976; *after* A. L. Bloom, *The Surface of the Earth*, Prentice-Hall, 1969.

This figure can be obtained from the data in Table 8-6, since oceans cover about 70 percent of the surface of the earth (0.70×125 cm/yr $+ 0.30 \times 41$ cm/yr $= 100$ cm/yr). As shown in Fig. 8-29, most evaporation occurs in the warm equatorial and tropical regions, particularly along latitude circles of about 20°N and 20°S. These are the belts of the semipermanent anticyclones within which subsidence carries warm, dry air downward.

Table 8-5 ESTIMATES OF TERRESTRIAL WATER

	Water volume (ℓ)	Percent of total water
Oceans	1.32×10^{21}	97.2
Continents		
Icecaps and glaciers	2.9×10^{19}	2.15
Ground water	8.3×10^{18}	0.62
Freshwater lakes	1.2×10^{17}	0.009
Saline lakes and inland seas	1.0×10^{17}	0.008
Vadose water (including soil moisture)	6.7×10^{16}	0.005
Average in stream channels	1.0×10^{15}	0.0001
Atmosphere	1.3×10^{16}	0.001
Totals	1.36×10^{21}	

Source: B. J. Skinner, *Earth Resources*, 2nd ed., Prentice-Hall, 1976.

Table 8-6   Annual Water Balance of the Oceans
and Continents in Centimeters per Year

	Evaporation (E)	Precipitation (P)	Total runoff (TR)
Oceans			
Atlantic Ocean	104	78	−26
Indian Ocean	138	101	−37
Pacific Ocean	114	121	7
Arctic Ocean	12	24	12
All Oceans	125	112	−13
Continents			
Europe	36	60	24
Asia	39	61	22
North America	40	67	27
(United States)	56	76	20
South America	86	135	49
Africa	51	67	16
Australia	41	47	6
Antarctica	0	3	3
All Continents	41	72	31

Source: W. Sellers, *Physical Climatology*, University of Chicago
Press, 1965.

The water vapor in the atmosphere is transported upward by turbulent motions and convective currents and carried long distances by the winds. When sufficiently strong and persistent updrafts exist, clouds and precipitation may occur. As a result, the water returns to the earth's surface. Since the average quantity of water vapor retained in the atmosphere changes little, the annual average precipitation over the earth must equal the evaporation. Therefore, the annual average rainfall equals about 100 cm, a quantity that greatly exceeds the depth of water vapor in the atmosphere at any time.

On the average, if all the water vapor in a vertical column of air extending from the ground to the top of the atmosphere were condensed, it would yield a layer of water about 3 cm deep. Since about 100 cm of precipitation falls in 365 days, the average planetary precipitation is about 0.27 cm/day. At this rate, the 3 cm of water vapor in the atmosphere would require 11 days to fall out. This is known as the *average turnover* or *residence* time of water vapor in the atmosphere. It indicates that water entering the atmosphere in one place is likely to fall out over regions far from its source, because over an 11-day period the winds will carry the water vapor over great distances.

The latitudinal distribution of precipitation over the entire earth is shown in Fig. 8-29. As expected, there is a pronounced peak over the equatorial regions where there are frequent thunderstorms associated with the intertropical convergence zone. Precipitation minima are associated with

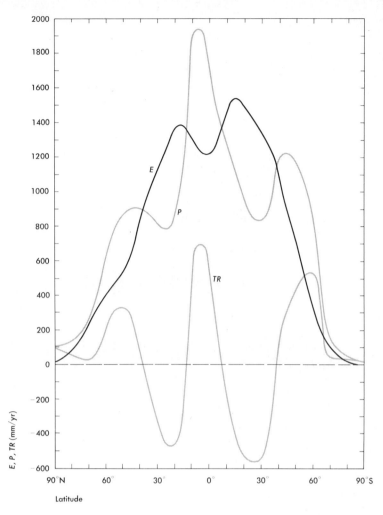

Figure 8-29 The average annual latitudinal distribution of evaporation (*E*), precipitation (*P*), and total runoff (*TR*). *From* W. D. Sellers, *Physical Climatology*, University of Chicago Press, 1965.

the sinking air in the subtropical anticyclones between latitudes 20° and 40°. Secondary precipitation maxima at middle latitudes are largely attributed to traveling cyclonic storms.

As seen in Table 8-6 and Fig. 8-28, more precipitation occurs over the oceans than over the land. Again recalling that the oceans occupy about 70 percent of the surface area of the earth, it is found that only about 22 percent of all rain and snow falls on the continents. Some of it wets the ground and vegetation and then evaporates again. Some of the water runs off the continents in rivers and streams and returns to the oceans. Also a small fraction percolates into the ground to recharge the aquifers and flow through the soil and gravel.

As already noted, the difference between precipitation and evaporation is called the *total runoff* (TR) and is shown in Fig. 8-29 and Table 8-6. A negative TR means that evaporation exceeds precipitation, and excesses in some places must supply the deficits in others in order to maintain a long-term balance. For example, over the Arctic Ocean precipitation exceeds evaporation and excess water flows into the Atlantic to balance the precipitation deficit. Over all the continents precipitation exceeds evaporation, and the total continental runoff balances the overall deficit of precipitation over the oceans. Note that the difference between 13 cm × 0.70 (fraction of ocean area) and 31 cm × 0.30 (fraction of land area) can be regarded as negligibly small when uncertainties in the estimates are taken into account.

It should be kept in mind that the quantities given in this section apply to averages over the earth as a whole. They imply a degree of consistency that is expected to occur over periods of several decades. As will be seen in a later chapter, over half a century there can be important changes in the earth's temperature accompanied by changes in the general circulation and in the quantity of water contained in sea ice and glaciers. At the same time, there would be vital changes in the hydrologic cycle. Since the supply of fresh water plays such an important role in human affairs, it is necessary to learn the degree to which the planetary water balance varies as the climate changes.

CHAPTER 9

Severe Storms

ACCORDING TO A WELL-KNOWN METEOROLOGICAL GLOSSARY,* a storm is "any disturbed state of the atmosphere, especially as affecting the earth's surface and strongly implying destructive or otherwise unpleasant weather." Contrary to the implications of this definition, most storms do more good than harm. They yield rain or snow to water growing vegetation and to fill lakes and reservoirs with fresh water needed for many human activities.

Unfortunately, certain storms can also be the sources of a great deal of misfortune. For example, an intense winter storm can bring very cold air, deep snow, and strong winds. When such blizzards occur, they can be disastrous. Grazing animals can become isolated and starve or freeze to death. Entire communities can be shut off from essential supplies.

Once or twice during the cold part of each year damaging ice storms can be expected over the regions of the United States and Canada bordering the Great Lakes and the St. Lawrence River. They occur in association with precipitation from warm fronts constituting parts of cyclonic systems. As snow from high altitudes falls through warm air above and in the frontal zone, the snow melts to form rain. If the cold air under the front has temperatures below 0°C, the resulting freezing rain can cause heavy loads of ice to collect on exposed objects and leave widespread damage in its wake. As was shown in Fig. 8-22, power lines and the poles holding them up are particularly susceptible.

Severe winter storms at least have the redeeming features that they develop gradually, move slowly, and can be predicted with a reasonable degree of accuracy. In most cases, steps can be taken to reduce the loss of life, especially human life.

On the other hand, severe storms of a convective nature develop so rapidly and are so small and short-lived that they are difficult to forecast. The best examples are violent hailstorms and tornadoes. They sometimes strike with frightening speed. In a matter of minutes a hailstorm may destroy a field of corn, and a tornado may demolish a score of buildings leaving wounded and dead in its path.

*Glossary of Meteorology, American Meteorological Society, Boston, 1959.

Still another type of severe storm is the hurricane and its counterparts in other parts of the world. These intense tropical vortices, though not containing the concentrated power of a tornado, are larger and longer-lived and therefore can do much more damage to life and property.

The threats posed to life and property in the United States by violent weather can be illustrated by the statistics presented in Table 9-1. Note that these are annual averages that have been rounded off. Individual storms can sometimes be more than five times more costly in lives and property. In other parts of the world extreme weather can take tolls that nearly defy belief. For example, a tropical cyclone which struck Bangladesh in November, 1970, took over a quarter of a million lives and rates as one of nature's most devastating calamities. The storm was a Southeast Asian counterpart of the South Atlantic hurricane.

Table 9-1   Storm Deaths and Damage in the United States

Type of storm	Approximate average annual deaths (1959–1975)	Approximate average annual property damage (1959–1975)
Lightning	198	$200 million*
Tornado	136	300 million
Hurricane	52	500 million
Hail	—	700 million†

*Includes livestock.

†Includes damage to food crops (about $600 million or about 1% of the cash value of all crops to the farmer).

Sources: E. Kessler, Director, National Severe Storms Laboratory, National Oceanic and Atmospheric Administration, Norman, Oklahoma, and H. M. Mogil and H. S. Groper, *Bulletin of American Meteorology Society*, 1977.

It may surprise some readers to learn that, on the average, lightning kills as many or more Americans than do hurricanes or tornadoes. Lightning fatalities occur mostly one or two at a time, scattered across the country, and do not get national attention unless the person struck is well known. When Lee Trevino, the famous golfer, was hit by lightning during a tournament, it made the headlines. Some of the subsequent stories pointed out that by taking simple precautions it is possible to avoid lightning hazards.

A common feature of the storms listed in Table 9-1 is that they involve atmospheric instability and deep convection. In this chapter we shall examine the properties of these severe storms, how they are observed, and how they are predicted.

THUNDERSTORMS

As was noted in Chapter 5, when the temperature decreases rapidly with height and the air is humid, the atmosphere is unstable. These conditions are most commonly encountered in maritime tropical air. When a sufficiently large volume of such air is given an initial upward velocity—by a perturbation in the windfield, an advancing front, or an orographic barrier—a convective current is set in motion. As the air rises, it accelerates as long as it is warmer than its environment. When condensation begins and the latent heat of vaporization is released, the added buoyancy will supply added acceleration to the updraft. Additional buoyancy is contributed by latent heat of fusion released as ice particles form. As the air continues ascending, the size of the cloud increases.

The vertical extent of a convective cloud depends on the temperature lapse rate, the humidity of the atmosphere, and the size of the rising volume of air. If there is a temperature inversion aloft, it may act as a stable barrier to prevent further convection. In some circumstances the updraft accelerates until it reaches the stable layer at the base of the stratosphere at altitudes as high as about 15 km. More commonly, thunderstorms extend to altitudes of about 10 km and often are topped by anvil clouds made up of ice crystals blowing out of and away from the main cloud regions (Fig. 9-1).

Although thunderstorms take on many sizes, shapes, and structures, they may be considered to fall into two broad categories: (1) local or air mass thunderstorms and (2) organized thunderstorms.

Local storms are fairly isolated storms having a short lifetime—less than an hour or so. They were studied in detail in the late 1940's by the Thunderstorm Project under the direction of the renowned meteorologist Horace R. Byers. On the basis of airplane penetrations, radar observations, and other measurements, it was proposed that such thunderstorms are made up of one or more "cells" that follow a three-stage life cycle represented schematically in Fig. 9-2.

Figure 9-1   A massive thunderstorm in Colorado. *Courtesy* National Center for Atmospheric Research, Boulder, Colorado.

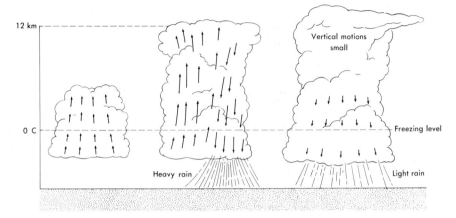

Figure 9-2   The cumulus, mature, and dissipating stages in the life of a single-cell thunderstorm. *Based on* H. R. Byers and R. R. Braham, Jr., *The Thunderstorm,* U. S. Government Printing Office, 1949.

In the cumulus stage the cloud is dominated by updrafts and contains growing cloud and precipitation particles. As the cloud enlarges, the updrafts become stronger and more widespread. In the upper parts of the cloud where the buoyancy is small and large quantities of liquid water and ice exist, a downdraft is initiated. It spreads downward and horizontally through the cloud. At maturity the cloud contains strong updrafts and downdrafts. During this period heavy rain occurs at the ground. In the final dissipating stage the cloud is characterized by weak downdrafts and light rain.

Under the cloud, starting even before rain reaches the ground, downdraft air encounters the ground and spreads rapidly outward. The air originating high in the cloud is cool and moist, and depending on the strength of the upper-level winds and the downdraft, out-flowing surface winds can be blowing at a high velocity and be very gusty. The cool air advances ahead of the thunderstorm, sometimes at speeds exceeding 30 m/s (67 mi/hr), and is capable of doing a great deal of damage to vegetation and buildings. More often the spreading downdraft air brings welcomed cooling on hot summer days.

As any airline pilot can attest, inside a thunderstorm, updrafts and downdrafts can be strong and turbulent. A Thunderstorm Project airplane flying through a storm at an elevation of about 5 km measured an updraft of about 25 m/s (56 mi/hr), but stronger draft speeds have been measured and calculated.

Although some small isolated thunderstorms, especially over the dry western United States, generate so little rain that it evaporates before reaching the ground, most local thunderstorms produce significant amounts of rain. Except for the occasional destructive out-blowing winds under the storm, lightning is the attribute of local thunderstorms posing a real danger.

THUNDERSTORM ELECTRICITY

As we have known since the days of Benjamin Franklin, lightning is a giant electrical discharge. Over the last couple of hundred years we have learned a great deal about thunderstorm electricity and lightning, but certain crucial questions still have no satisfactory answers. It is now well accepted, as shown schematically in Fig. 9-3, that the upper part of a thunderstorm is predominantly positive and the lower part is mostly negative. A smaller, positively charged center is often found in the rain region near the ground.

There is still much debate among the experts about the principal mechanism in thunderstorm electrification. Most authorities, on the basis of laboratory experiments and field observations, believe that electric charges are generated and separated in the process of precipitation growth and fall. It has been shown that when ice pellets fall through a region of smaller cloud droplets and raindrops, the collisions, and in the case of supercooled droplets, the freezing of water on the ice pellet can cause electric charge separation. The falling ice pellet acquires a negative electric charge, while rebounding cloud particles or fragments of the impacted water drops are left with positive charges. Because their terminal velocities are relatively small, these particles are separated from the ice pellets. In an updraft the positively charged particles are carried toward the top of the cloud. The positive charge in the rain has been explained as a result of the acquisition of positive charges released from the ground.

A few atmospheric scientists argue that the electrification of thunderstorms does not require the presence of ice particles. They cite a number of observations of lightning in convective clouds having temperatures above freezing. These authorities propose that the charge in a thunderstorm is separated by the selective capture and transport by cloud droplets of tiny positive and negative ions in the atmosphere.

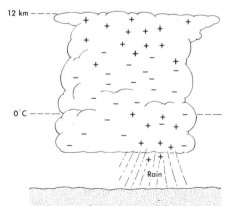

Figure 9-3 Schematic representation of the electric charge distribution in a mature thunderstorm.

As the electric charge accumulates in the cloud, the *electrical field* strength between the cloud and ground increases. This quantity is a measure of the gradient of electrical potential between two points. For example, if you had two horizontal metal plates 2 m apart and the upper one had a positive potential of 5 V and the bottom plate had a potential of -5 V, the electric field between the plates would be 10 V/2 m $=$ 5 V/m.

In fact, a situation somewhat analogous to this example exists on earth. In Chapter 2 we mentioned the presence of the ionosphere in the upper atmosphere. The ionosphere represents a vast layer of positive ions. It can be considered to act as if it were a spherical plate charged positively while the ground is charged negatively. The presence of the ionosphere accounts for the fact that during days with clean air and cloudless skies the vertical electric field measured at the ground amounts to about 100 V/m.

When clouds occur, the droplets and ice crystals accumulate electric charge and the electric field between the ground and the cloud increases. In the case of a potential thunderstorm, the field strength can increase to very large values over periods of minutes. At the ground, under thunderstorms just before lightning occurs, fields have been found to be as high as 10,000 V/m. High as this value is, it still is not enough to initiate a lightning discharge. In order for an electrical arc to jump across two conductors in dry air, the field strength between them must exceed the *breakdown potential* about 3×10^6 V/m. Smaller fields can lead to spark discharges in clouds of water droplets or ice crystals.

When the breakdown potential is reached, a lightning flash is initiated. It may consist of a single stroke lasting a hundredth of a second or it may consist of as many as 30 or more strokes in quick succession over a period of about a second. More will be said in the next chapter about the nature of lightning and the associated thunder.

ORGANIZED THUNDERSTORMS

Most often the storms regarded as being in the organized class are those that form in lines or bands of thunderstorms, sometimes called *squall lines* (Fig. 9-4). They commonly are initiated along a cold front, or ahead of, and nearly parallel to it. The lines of thunderstorms travel through the warm sector of a cyclone at speeds often exceeding the speed of the cold front. Large thunderstorm systems of this type can last for many hours with relatively slow changes in their properties. As the storms move across the country, they may deposit swaths of hail or tornadoes.

In Fig. 8-23 it was shown that thunderstorms occur most frequently over Florida with a secondary frequency maximum over New Mexico. The same illustration shows that hail is most common over eastern Wyoming and

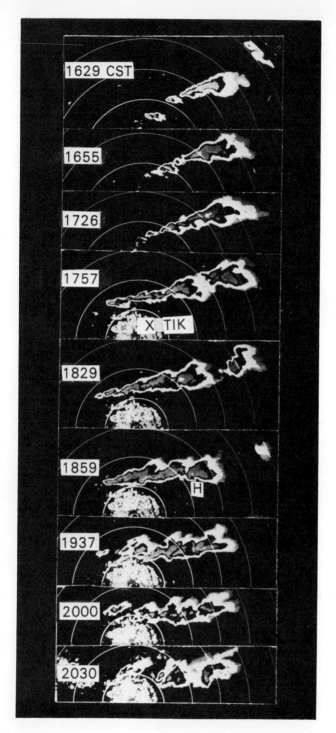

Figure 9-4 Radar observations of a line of thunderstorms moving southward toward Norman, Oklahoma, which is located at the center of the circular markers. The markers are 37 km (20 nautical miles) apart. Therefore, at 1829 CST the line was about 110 km long. *From* E. Kessler, *Weatherwise*, 1970, **23**: 56–69.

Colorado. Figure 9-5 depicts the distribution of tornadoes over the United States. Most of them occur over the southern Great Plains and the Middle West. These data show that the intense, organized storms capable of yielding hail and tornadoes are products of conditions that occur most often to the east of the Rockies, the giant barriers which constrain the humid, unstable air from the Gulf of Mexico from moving westward. Before examining the conditions that specifically favor tornadoes, we should recall that, as noted in Fig. 8-27, large hail forms in thunderstorms having strong persistent updrafts that are tilted into the wind. The winds in the storm's environment increase with height. Air containing water and ice particles rises through the cloud and much of it flows out the top in a forward direction. Such storms may exhibit anvil clouds extending downwind to very great distances.

Typically, hailstorms are not steady and continuous producers of hail. Instead, it is more common to have them yield showers of hail that are laid out in the form of streaks along the path of travel. They may be from a few hundred meters to many kilometers long. On the average, a hail streak is about 800 m wide and 8 km long, but some can be much longer. On some unusual occasions a hailstorm may persist over a single location and yield an incredible quantity of ice in a brief period. For example, at Selden in

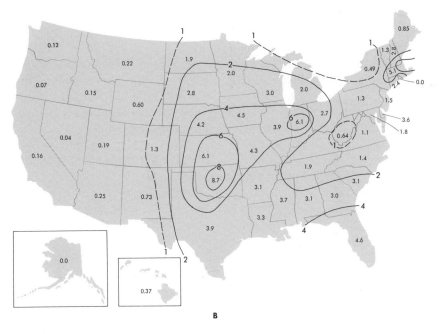

B

Figure 9-5 Average annual number of tornadoes per 25,000 square kilometers based on state averages for the years 1953–1970. See Fig. 8-23 for maps of thunderstorm and hailstorm frequency. *Source:* National Weather Service.

northwest Kansas on June 3, 1959, hail fell for 85 min over an area 10 km by 15 km and covered it with a layer of hailstones 46 cm (18 in.) deep.

Hailstorms can occur at almost anytime of the year if the appropriate meteorological conditions occur, but they are predominantly spring and early summer phenomena. May is the month of highest hail frequency. Most storms occur during the hours 2:00 P.M. to 9:00 P.M. local time with the highest frequency between 4:00 P.M. and 7:00 P.M. As will be seen in the next section, tornadoes exhibit much the same daily and annual patterns of occurrence.

TORNADOES

Because of their concentrated power, tornadoes are probably the most feared of all weather phenomena. A tornado usually strikes suddenly, with little warning, and in a few minutes causes extensive damage to property, injuries, and loss of life. As shown in Fig. 9-6, a tornado may have the appearance of a narrow funnel, cylinder, or rope extending from the base of a thunderstorm to the ground. The visible funnel consists mostly of water droplets formed by condensation in the funnel. Near the ground blowing dust, leaves, and other debris identify the presence of a strong vortex. Tornadoes and weak visible vortices occurring over water are called *waterspouts*. They are most common over tropical and subtropical oceans.

Tornadoes are generally small, typically less than a few hundred meters in diameter, but some are larger than a kilometer. With rare exceptions the winds are cyclonic, i.e., they blow counterclockwise in the Northern Hemisphere. The funnels usually touch the ground for only a few minutes or so, but some have been reported to last for much more than an hour. Maximum wind speeds are estimated to reach about 100 m/s (224 mi/hr). At one time it was thought, on the basis of the damage surveys immediately after the storm, that tornado wind speeds were much higher, possibly even reaching the speed of sound.

Waterspouts generally range in diameter from a few meters to a few hundred meters and have wind speeds estimated to be from as low as 20 m/s (45 mi/hr) to more than 60 m/s (134 mi/hr).

Most of the fatalities and damage are caused by a small number of large, long-lived tornadoes. According to two experts in the United States, Allen Pearson and T. T. Fujita, during the decade ending in 1970 fewer than 2 percent of the largest tornadoes accounted for 85 percent of the fatalities. Pearson refers to the strongest 5 percent to 10 percent of the storms in the United States (about 50 per year) as *maxi-tornadoes.*

Tornadoes are almost always associated with severe thunderstorms. As a result of the interactions of the thunderstorm and its environment, particularly as cool downdraft air strikes the ground, a small-scale cyclonic circula-

Figure 9-6 Various types of tornadoes: (Top, left) *Courtesy* Bill Males; (Top, right) *Courtesy* National Oceanic and Atmospheric Administration; (Bottom) a waterspout, *courtesy* J. H. Golden, National Oceanic and Atmospheric Administration.

tion is established in the vicinity of a tornado-producing thunderstorm. Such a *tornado cyclone*, perhaps 5 km or so in diameters, is most often located on the right-hand side of a storm, that is, on the south side of a thunderstorm moving toward the east. One or more tornado funnels may form within the tornado cyclone as it moves with the thunderstorm with which it is associated (see Fig. 9-7). The translation speed averages about 15 m/s (34 mi/hr), but it varies from nearly 0 m/s to more than 30 m/s in rare cases.

In 1970 Fujita concluded that within a tornado there are still smaller, intense whirls that he calls *suction vortices*. They might have diameters of about 10 m. A small, short-lived tornado might have only one; a maxitornado would have many suction vortices. Fujita proposed that these small vortices can account for occasional observations of virtually total destruction of one structure while another one 10 m away is left unscathed.

The pressure in a tornado funnel is substantially lower than the surrounding atmospheric pressure. Figure 9-8 shows the changes of pressure when a tornado passed near a barograph. Such measurements are scarce, but it has been estimated that in a severe tornado the central pressure might be more than 100 mb less than the pressure in the environment.

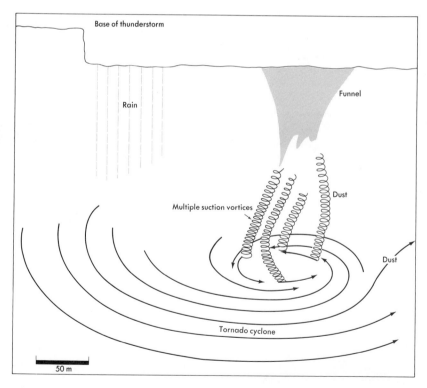

Figure 9-7 Schematic representation of tornado cyclone, tornado funnel, and suction vortices. *Courtesy* T. T. Fujita, University of Chicago.

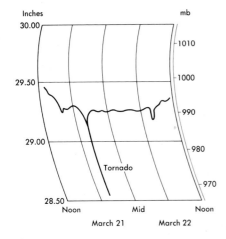

Figure 9-8 The characteristic low pressure at the center of a tornado shows up in this record of a storm that passed Dyersburg, Tennessee, at 2050 CST on March 21, 1952. *From* J. A. Carr, *Monthly Weather Review*, 1952, **80**: 50–58.

The low pressure and strong winds account for the destructive nature of a tornado. When one of them moves over a building, the outside pressure suddenly drops while the pressure inside changes slowly, particularly if the windows and doors are closed. As strong winds blow over a building, dynamical effects tend to produce high pressures on the upwind side and low pressures on the lee side of the structure. When the pressure inside a building substantially exceeds the pressure on the outside, the result is a strong pressure force that can cause the roof and walls to blow outward with explosive violence (Fig. 9-9). For example, if the pressure inside a house is 1 lb/in² (69 mb) greater than the pressure outside, the force on a ceiling 10 ft by 10 ft is 14,400 lb, equivalent to the weight of a mass of 6545 kg. This would be large enough to do extensive damage. In some cases, the force might pick up the roof of the building.

Exceedingly heavy objects can be carried away by the powerful tornadic winds. Railroad cars and houses have been moved; remnants of demolished buildings have been scattered over long distances.

An interesting aspect of tornadoes is the loud, distinctive noises they produce. People who have had the misfortune to have one pass near by have reported sounds resembling "a thousand railway trains," "the buzzing of a million bees," or "the roar of flights of jet airplanes." An adequate explanation still has not been obtained.

Although, as already noted, tornadoes are produced by thunderstorms, meteorologists still do not agree on the mechanisms that cause them. As shown in Table 3-1, the energy of a tornado is about a hundredth of that of a typical thunderstorm. For this reason, it is reasonable to suppose that the thunderstorm causes the funnel and supplies its energy. It has been speculated that electrical discharges might initiate and maintain the funnel, but this idea has few supporters. Nevertheless, there is evidence that at least some tornadoes are electrically active.

Figure 9-9 Damaged houses in Rochester, Indiana, caused by a tornado on April 3, 1974. *Courtesy* U. F. Koehler, Ball State University, Muncie, Indiana.

An attractive theory for explaining tornadoes is that they are initiated in a strong downdraft when the descending air has strong rotational characteristics imposed by the wind pattern around the thunderstorm. It tends to account at least in part for radar observations showing vortices aloft, within thunderstorms, some tens of minutes before the appearance of the funnel at the ground. Also, visual observations indicate that the visible tornado funnel is first seen just under the cloud base and that it works its way downward.

Although tornadoes occur in many other countries (Fig. 9-10), the United States experiences by far the highest frequency. In an average year there may be perhaps 700 separate tornadoes reported. Commonly, a single large thunderstorm system will produce several tornadoes. In an intense outbreak as many as 30 to 40 separate funnels will be generated as the parent storm travels over distances of several hundred kilometers. The most extensive outbreak of storms during this century occurred on April 3 and 4, 1974. Over a 24-hour period 148 tornado funnels were observed over a major fraction of the eastern half of the United States. According to official reports, these storms led to 315 fatalities, 6142 injuries, and damage to property estimated at $600 million. Clearly, these data are extreme when compared with the statistics in Table 9-1.

194

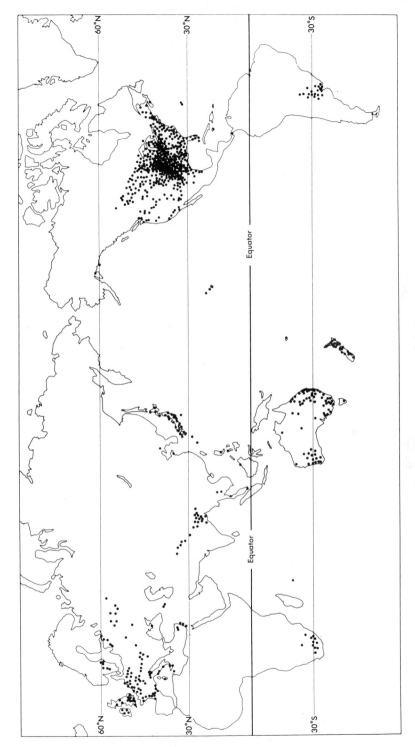

Figure 9-10 Tornado occurrences around the world expected over a 4-year period. *From* T. T. Fujita, *Weatherwise*, 1973, **26**: 56–62.

Until the late 1940's the U.S. Weather Bureau did not issue tornado forecasts. It was recognized that tornadoes were difficult to predict, and there was concern that the forecasts would frighten the public, often unnecessarily. Over the succeeding years there have been major impovements in methods for forecasting storms, detecting them, and communicating tornado watches and warnings to the public.

Tornadoes are most frequent in the late afternoon and early evening and occur most often in the spring and early summer months. It is during this period that the necessary meteorological conditions are most likely to be present. Figure 9-11 illustrates circumstances in which tornadoes are common. Moist tropical air from the Gulf of Mexico sweeps northward over the southern Great Plains. Aloft, a strong current of dry air from over the desert Southwest flows over the moist air. The advancing frontal system initiates the lifting of the moist air and leads to convection. In the warm sector of the cyclone, thunderstorms are a common event. If the air is very unstable, an intense squall line with tornadoes will be likely to develop.

Tornado forecasters are confronted with several major problems. An inadequate understanding of the nature of tornadoes and the processes of development and dissipation certainly contribute to the difficulties. Even if we knew much more about severe thunderstorms and the sequence of events leading to tornado occurrence, the spacing of observations in distance and time would put limits on tornado forecasting. The rawinsonde stations used to observe the temperature, humidity, and wind structure of the atmosphere are separated by several hundred kilometers and usually take observations only twice a day. It is not surprising that these observations lead to forecasts that can do no better than to call for tornadoes to occur over areas of perhaps 65,000 km² (about 25,000 mi²).

When tornadoes are expected to occur over the next several hours, National Weather Service forecasters issue a *tornado watch*. It encompasses a large number of people, only a small fraction of whom are likely to be affected even if tornadoes do occur. Those within the area covered by the tornado watch should be alert to the possibility of violent weather. A plan of action should be formulated in the event a tornado occurs. A local radio station should be monitored preferably by means of a battery-powered radio.

If a tornado is observed, the National Weather Service issues a *tornado warning* and advisories are broadcast. Unfortunately, tornadoes usually strike suddenly and swiftly and the time for evasive action is brief.

Notwithstanding their limitations, the existing forecasts are of substantial value because they focus attention on those regions of greatest vulnerability. In order to locate tornadoes more precisely, it is necessary to use visual observations and remote detection devices such as radar, weather satellites, or radio receivers that detect the radio signals emitted by the storms.

It is unfortunately true that even in these years of space exploration, most tornadoes are detected visually by ordinary citizens who spot the funnels

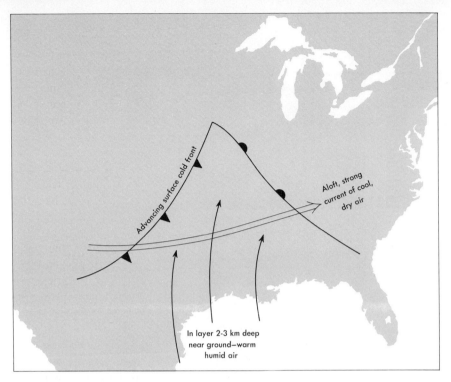

Figure 9-11 Meteorological conditions favoring the formation of severe thunderstorms and tornadoes over the southern Great Plains.

and notify local officials. It has been proposed that observations of the behavior of thunderstorms from satellites can indicate regions, but these techniques are not yet ready for application.

Since World War II radar has been used to observe the location, size, movement, and intensity of thunderstorms. It has been employed extensively in research and day-to-day weather operations. A network of radars throughout the United States monitors the formation and behavior of thunderstorms and other rain and snow storms. A conventional radar of the type used by the National Weather Service detects the presence of water drops and ice particles. In general, the greater the size and concentration of the precipitation particles, the more intense the radar echo. For this reason, radar can be used to measure the intensity of rain and snow.

Since a conventional radar cannot specifically identify a tornado, it is necessary to use characteristics of the echoes to infer the presence of a tornade. It has been found that long-lasting thunderstorms that extend to great altitudes and produce intense radar echoes are likely to have associated violent weather—hail and tornadoes. When a weather forecaster observes such a thunderstorm echo moving across a region within which tornadoes have been predicted, he can localize the areas where tornadoes are most likely to occur. If possible, attempts are made, via radio or telephone, to obtain visual observations from state police or other individuals in the vicinity.

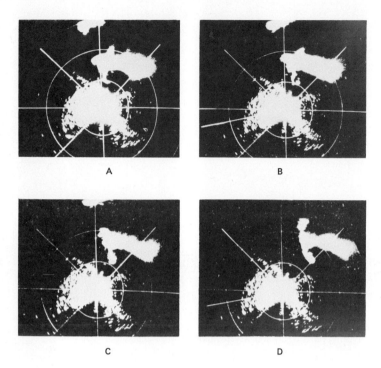

A

B

C

D

Figure 9-12 A tornado-producing thunderstorm detected by radar at times (A) 1713, (B) 1716, (C) 1719, and (D) 1725. The tornadic activity was associated with the hook-shaped protuberance extending from the southwest part of the large thunderstorm echo. The circular markers are 16 km apart. *From* G. E. Stout and F. A. Huff, *Bulletin of American Meteorological Society*, 1953, **34**: 281–284.

Occasionally, the shape of a radar echo is a good indication of the presence of, or the impending formation of, tornadoes. The most reliable such indicator is a hook-shaped echo extending from a large thunderstorm echo (Fig. 9-12). Unfortunately, such distinctive indicators do not occur often enough. Most of the time, the shape of the echo is of little use in identifying tornadoes.

One of the greatest values of radar is to track tornado-producing thunderstorms. As noted earlier, some traveling storms produce a series of funnels. Once the *mother storm* is identified, it can be followed, and people in its path can be given warnings of its approach, perhaps an hour or two in advance of the arrival of the storm.

A *Doppler radar* can observe the same quantities as can the conventional radars just discussed, but in addition it can measure the speed, toward or away from the radar, of targets such as raindrops or other liquid or solid objects. Since a characteristic feature of a tornado is the high wind speed, it is reasonable to expect that a Doppler radar would be well suited for tornado detection. Research carried out in the middle 1970's has indicated that an appropriately designed Doppler radar can identify and locate tornadoes. In some cases, it is possible to detect evidence of tornado development as much as 20 min before the funnel reaches the ground.

HURRICANES

The name hurricane is given to tropical cyclones occurring over the Atlantic or the eastern North Pacific Oceans and having maximum wind speeds greater than 32.6 m/s (73 mi/hr). The same type of storm in the western North Pacific is called a *typhoon* and in the Indian Ocean it is called a *cyclone*. The intense tropical storms off the west coast of Mexico have been named ciclóns by the natives of that country. Various other names are used in other parts of the world. Figure 9-13 shows areas where they are observed and the tracks they commonly follow. For convenience, we shall call all of them hurricanes.

As is the case with tornadoes, the chief characteristics of hurricanes are the low central pressures and the high wind speeds. But the two storms differ greatly in size and duration. A typical hurricane is a nearly circular vortex some 500 km in diameter and it lasts for many days. Some storms last for more than a week.

The central pressure in a strong hurricane may be more than 50 mb lower than the pressure at the outskirts of the vortex, and in extreme cases the pressure difference could approach 100 mb. As a result, a hurricane is characterized by steep pressure gradients (Figs. 9-14 and 9-15) and high wind speeds (see Fig. 12-9). Peak winds, sometimes exceeding 80 m/s (179 mi/hr), occur usually within about 30 km of the storm center.

In the innermost portion of the storm the winds are light and the sky contains few clouds. This is called the *eye* of the storm. Its diameter averages about 20 km, but in some large storms the eye may be more than 40 km in diameter.

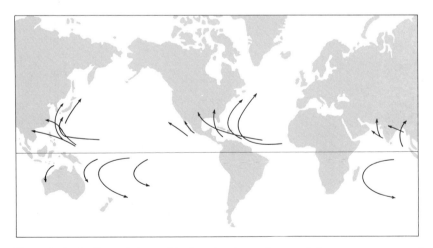

Figure 9-13   Regions of the world where hurricanes form and the most common tracks they follow. *From* G. E. Dunn and B. I. Miller, *Atlantic Hurricanes*, Louisiana State University Press, 1960.

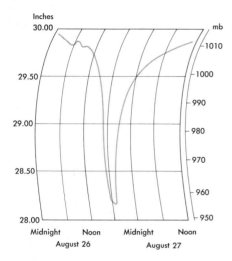

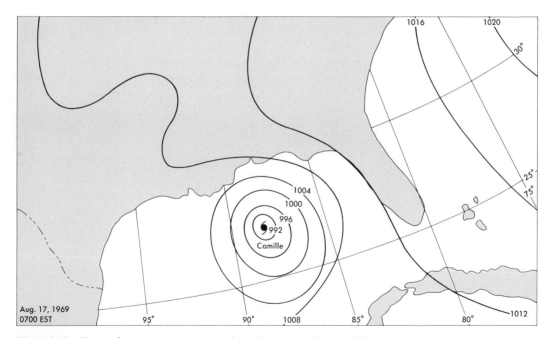

Figure 9-14 The changes in pressure (in inches of mercury) as a hurricane passed West Palm Beach, Florida, in August, 1949. (Note that 1 in. Hg = 33.9 mb = 0.49 lb/in²). *From* R. T. Zoch, *Monthly Weather Review*, 1949, **77**: 339–341.

Figure 9-15 The surface pressure pattern of hurricane Camille at 0700 EST on 17 August 1969. The eye of this hurricane passed over the coast of Mississippi at about 2300 EST on the same day. Fig. 9-18 shows the path it followed from August 15 to 20, 1969.

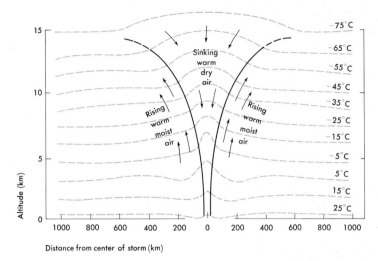

Figure 9-16   Schematic cross section showing temperature distribution and air motions in a mature hurricane. The heavy solid lines show the boundaries of the eye. *Based on a model by* E. Palmén.

A schematic cross section through a hurricane is shown in Fig. 9-16. The air in the interior part of the storm sinks and is drier and warmer than the air outside the storm vortex. This accounts for the description of a hurricane as a warm core cyclone. Strongest upward velocities are found outside the ring of peak winds, in the cores of convective clouds and thunderstorms. They are not symmetrically distributed around the storm. Instead, there usually are spiral bands of precipitation with the heaviest concentrations in the right-forward quadrant of the hurricane. In this stormy region tornadoes are likely to develop. The pattern of clouds and precipitation stands out clearly on radar scopes (Fig. 9-17).

Hurricanes develop over warm oceans and derive most of their energy from the underlying water. Latent heat is transported into the storms and released in the process of cloud and rain formation. The precise mechanism by which hurricanes are formed still is not clear. What is known is that there are many pressure disturbances over the tropical oceans and that a small percentage become hurricanes. For example, over the Atlantic Ocean in an average year there are about 100 rain storms detected by means of weather satellites. When the winds in these weak storms are less than 17.5 m/s (39 mi/hr), they are called *tropical depressions*. Some of them originate as low-pressure systems that formed over the African continent and propagate westward.

In an average year, mostly during the months of August, September, and October, some ten depressions intensify and are given the name *tropical storms*. This means that peak winds are between 17.5 m/s and 32.6 m/s (39 mi/hr to 73 mi/hr). Six of these storms reach hurricane strength and two cross a United States coastline. In fact, the number of Atlantic hurricanes varies greatly from year to year, sometimes being more than ten. Through the

Figure 9-17 Hurricane Donna at 0730 EST on September 10, 1960, seen on a radar at Miami, Florida. The bright spiral bands represent regions of rainfall. The circular marker is at a range of 100 nautical miles (185 km). *Courtesy* L. F. Conover, National Hurricane Research Laboratory, Miami, Florida.

systematic satellite surveillance of the earth it has been observed that the number of tropical storms and hurricanes is greater than was once suspected.

As suggested in Fig. 9-13, the most common paths followed by hurricanes are determined by the prevailing wind patterns. The storms are transported by the easterly trade winds and then they curve poleward as they approach the continents. As you might suspect, this is an oversimplification because the wind patterns sometimes differ markedly from the averages represented by the general circulation of the atmosphere. A hurricane is steered by the major current of air in which it is located while, at the same time, to a lesser extent, moving through the stream. As the current changes, the hurricane path also changes. Figure 9-18 shows actual storm tracks of selected hurricanes over the western North Atlantic Ocean. It is evident that in some cases, the storm track can change abruptly.

As hurricanes move over land or over higher-latitude oceans, they become weaker. This occurs primarily because the energy input is reduced as the storm moves away from regions of warm ocean water. In addition, when a hurricane passes over a continent, the terrain exerts additional frictional forces that act to reduce the wind speeds.

As noted early in this chapter, hurricanes passing over land can be very destructive and lethal. A single storm, such as hurricane Camille, which swept into Mississippi in August, 1969, did about $1.5 billion dollars of damage. Torrential rains from hurricane Agnes in 1972 did twice as much damage over the eastern United States. Although it was not the case with Agnes, the main cause of damage and loss of life is usually the *storm surge*, a wave of ocean water generated by the hurricane winds as the storm approaches a coast. A wall of water some 3 m or more in height can be produced. It can sweep over low-lying land, and coupled with heavy rains, it can cause flooding on a massive scale (Fig. 9-19). Evacuation to high ground is the only sure means of survival.

Hurricane winds over the open oceans can cause waves of spectacular heights. They depend on the strength of the wind, the size of the storm, and its duration. In the Atlantic, an average hurricane produces waves some 10 m to 12 m high. Waves generated in different quadrants of the storm cross one another, creating wave peaks that may be higher than 20 m in extreme cases. Such waves were reported in 1935 by a Japanese naval vessel that inadvertently sailed into a typhoon. As one can imagine, the waves can toss around even the largest ship and make sailing through a hurricane a terrifying experience.

The waves produced by a hurricane propagate outward in all directions. Those initiated in the right-hand side of the storm move rapidly in the direction of the storm's movement. A typical hurricane may travel at a speed of about 6 m/s (13 mi/hr) and advance about 500 km in a day. The storm-generated waves propagate much faster, covering a distance of about 1000 km to 1500 km in a day. As the waves leave the storm region, their amplitudes

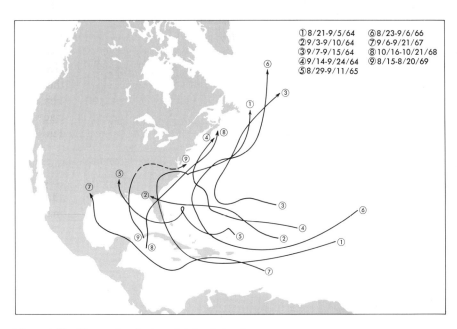

①8/21-9/5/64	⑥8/23-9/6/66
②9/3-9/10/64	⑦9/6-9/21/67
③9/7-9/15/64	⑧10/16-10/21/68
④9/14-9/24/64	⑨8/15-8/20/69
⑤8/29-9/11/65	

Figure 9-18   The tracks of selected Atlantic hurricanes.

Figure 9-19 Flooding in New Orleans caused by hurricane Betsy on September 9 and 10, 1965. American Red Cross photo.

diminish, the crests take on a flatter shape, and the waves are called *swells*. Before the days of orbiting satellites, radar, and airplanes, the arrival of swells was often the only warning to sailors and coastal dwellers that a hurricane was imminent.

The strong winds in a hurricane do a great deal of destruction to man-made structures and vegetation. Huge trees can be blown over and loose objects carried away. Wind damage is most severe near coastlines because the peak wind speed near the ground decreases as the storm passes over land. These problems are further aggravated by the presence of tornadoes that often are produced by the thunderstorms in the hurricane. The relatively high frequency of tornadoes shown in Fig. 9-5 to occur in Florida is explained by the fact that Florida has been in the path of many hurricanes.

Figure 9-20 Satellite observations of hurricane Belle off the east coast of the United States on August 8, 1976. *Courtesy* National Environment Satellite Service, National Oceanic and Atmospheric Administration.

Fortunately, by means of weather satellites, hurricanes can be detected early in their development and tracked throughout their existence. Figure 9-20 shows how a weather satellite observed hurricane Belle on August 8, 1976, when it was located off the east coast of the United States. The edge of the storm and the spiral rain bands are clearly evident.

When a hurricane is within a couple of hundred kilometers of land, high-powered radars follow the storm's development and movement. Specially equipped airplanes can be used to make measurements of the storm's position as well as its pressure and wind patterns. Modern forecasting and communication techniques make it possible to warn people in areas likely to be affected in sufficient time for them to take steps needed to reduce property damage and insure their own safety. It is particularly vital that people in low-lying coastal areas move to higher inland locations when a hurricane is expected. More will be said about hurricanes in Chapter 12 in which there is a discussion of the attempts being made to weaken them by means of cloud seeding.

CHAPTER 10

Atmospheric Optics and Acoustics

EVEN IF A PERSON KNEW NOTHING ABOUT THE ATMOSPHERIC SCIENCES, the sights and sounds of the atmosphere would be enough to stir the imagination. Everyone is familiar with lightning and thunder and the occasional rainbow, but there are many other visual and auditory experiences that are not so common. In some parts of the world the mirage is the source of wonder; in others the appearance of beautiful mother-of-pearl clouds or an aurora such as the one in Fig. 2-14 lift eyes to the sky in admiration. Students of atmospheric optics observe a large number of interesting phenomena as light from the sun or the moon passes through clouds of water droplets or ice crystals. In this book we discuss only a few of the most frequent optical effects. A reader who has a special interest in this subject is referred to the fascinating book by M. Minnaert entitled *The Nature of Light and Color in the Open Air.**

RAINBOWS

As noted in Chapter 2, sunlight is made up of a mixture of wavelengths ranging from red to violet. A prism oriented as shown in Fig. 10-1 will cause the colors of the sun's rays to be separated. The dispersion of the color spectrum can be explained in terms of a phenomenon known as *refraction*. It comes about because the speed of an electromagnetic wave depends on the property of the medium through which it passes.

Figure 10-2 illustrates the passage of light from one medium, perhaps water, into another medium, perhaps air. Because of the differing physical properties, the wave speed is greater in air than it is in water. As a wave front emerges from the water, it is reoriented because the part in the air moves faster than the part in the water.

The bending of light rays is easily demonstrated. Take a pencil and dip part of it in water (see Fig. 10-3). When you look down at it from the side, the pencil will appear shorter than it actually is. This happens because the light rays being reflected from the pencil are deflected at the surface of the water.

*Dover Publications, Inc., 1954.

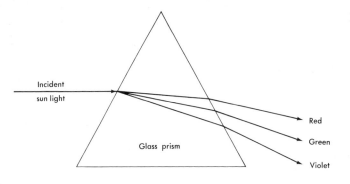

Figure 10-1 As sunlight passes through a glass prism, refraction occurs. Since the short waves (violet and blue) are refracted more than the long waves (red), there is a separation of the various colors.

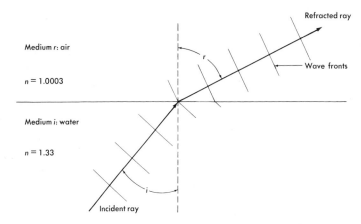

Figure 10-2 Light waves are refracted as they pass from one medium to another because the speed of the wave increases as the refractive index of the medium decreases.

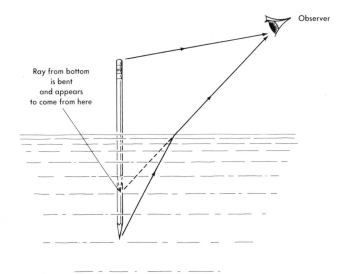

Figure 10-3 The ray of light from the bottom of the pencil changes direction as it emerges from the water. As a result, the pencil looks shorter than it actually is. The bending of the light ray occurs because the refractive index of the air is less than that of water.

The velocity of light in any medium is given by the speed of light in a vacuum (3×10^8 m/s) divided by a quantity called the *index of refraction*. Table 10-1 gives the refractive indices of various substances and the corresponding speeds of light through them. It can be seen, for example, that light passes through water at a velocity of 2.26×10^8 m/s, a value substantially less than the speed of light in air.

Table 10-1   Indices of Refraction (n) of Various Substances*
                 for Visible Light

Medium	n	Speed of light† ($\times 10^8$ m/s)
Vacuum	1.0000	3.000
Air (lower atmosphere)	1.0003	2.999
Ice	1.31	2.29
Water	1.33	2.26
Olive oil	1.47	2.04
Flint glass	1.61	1.86
Diamond	2.42	1.24

*The values in this table apply specifically to a wavelength of 0.589 μm (yellow) but they are approximately correct for all visible radiation.

†Speed $v = c/n$ where $c = 3.000 \times 10^8$ m/s.

As you would expect, the degree of refraction as a light ray moves from one medium to the next depends on the difference in indices of refraction. The change of direction of a light ray is given by Snell's law of refraction.* It states that the greater the ratio of the indices of refraction, the greater the bending of the ray as it moves from one medium to the next.

The value of the refractive index depends not only on the properties of the medium but also on the wavelength of the electromagnetic radiation. Over the visible part of the spectrum the variations are small, as shown in Table 10-2. Nevertheless, they are important because they lead to the dispersion of white light into its various color components as it passes through a glass prism, or a raindrop, or an ice crystal. Since the refractive indices at the violet end of the spectrum are larger than those at the red end, according to Snell's law, the violets and blues are refracted more than are the yellows and reds.

*Snell's law of refraction can be expressed as follows:

$$\frac{\text{sine } r}{\text{sine } i} = \frac{n_i}{n_r}$$

where, as shown in Fig. 10-2, i is the angle of incidence, r is the angle of refraction, and n_i and n_r are the refractive indices of the two media. The value of the sine of any angle can be obtained from a table of trigonometric functions.

Table 10-2 INDICES OF REFRACTION OF VARIOUS SUBSTANCES
FOR VISIBLE WAVELENGTHS

Wavelength (µm)	Color	Ice (T = −3°C)	Water (T = 0°C)	Zinc crown glass
0.434	Violet	1.316	1.341	1.528
0.486	Blue	1.313	1.338	1.523
0.589	Yellow	1.309	1.334	1.517
0.656	Red	1.307	1.332	1.514

Sources: W. E. Forsythe, *Smithsonian Physical Tables*, 9th ed., 1954, and
E. N. Dorsey, *Properties of Ordinary Water-Substance*, 1940.

Figure 10-4 shows how a raindrop reflects and refracts sunlight. As the light waves pass into the water, they are refracted and the separation of colors is started. At the back surface of the drop the rays are partially reflected, although most of the incident energy passes through the back surface of the raindrop. As the reflected waves leave the drop, they are refracted again. By this time the angular separation of different colors has reached its maximum. The spacing between the red and violet in Fig. 10-4 has been exaggerated to make clear what happens. In fact, the difference in angle between the red and violet rays emerging from a raindrop is small, being less than 2°. The angle between the incoming sun's ray and the outgoing colored rays is close to 42° for all colors.

The color of the light, from any particular drop, which reaches an observer's eye depends on the angle made by the sun's rays and the position

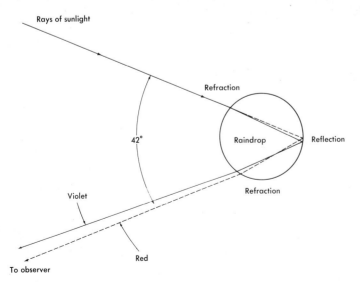

Figure 10-4 The path of a ray of sunlight as it travels through a raindrop. The light is refracted as it enters and leaves the raindrop. As in the case of the prism (Fig. 10-1), refraction causes a separation of white light into different colors.

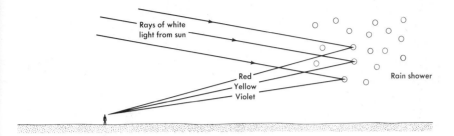

Figure 10-5 The primary rainbow exhibits violet and blue on the inside of the bow and the red on the outside. This occurs because the short waves are refracted more than the long waves.

of the observer. This point is illustrated in Fig. 10-5. Each of the raindrops in the shower separates the sunlight into its various colors ranging from red to yellow to violet, but the observer sees only the red from certain drops, only yellow from drops at a lower altitude, only blue from still lower drops, and only violet from the still lower ones.

Figure 10-5 shows that the rainbow always appears in the opposite side of the sky from the sun. You see it when the sun is at your back as you look at a distant shower. The diagram also shows that red appears on the outside of the bow. The colors progress to orange, yellow, green, blue, and violet as you go toward the inside of the rainbow. Because of the geometry involved, when rainbows are viewed from the ground, they appear as semicircles. This encouraged stories about the "pot of gold at the end of the rainbow." If you are ever fortunate enough to fly over a region of raindrops when the sun is high in the sky, you will have the thrill of looking down and seeing a rainbow in the form of a complete ring.

The *primary rainbow*, which we have just described, comes about because of a single reflection off the back surface of the raindrops in the manner illustrated in Fig. 10-4. On some occasions, depending on the size of the rain region and the position of the sun, you can see a *secondary rainbow*. It occurs as a result of two reflections from the back of the raindrops in the manner illustrated in Fig. 10-6. Note that because the red is refracted less than the violet, the arrangement of colors in the secondary bow is reversed from that of the primary bow. The red is on the inside, and the violet is on the outside of the bow.

As shown in Fig. 10-6, because of the geometry of the reflections, the angle between the sun's rays and those reaching an observer forms an angle of 50°, some 8° greater than the angle subtended by the primary rainbow. Because there are two reflections and each one returns only part of the incident energy, secondary bows are weaker in intensity than the primaries.

Occasionally, when the moon is bright, its light can be adequately reflected and refracted to produce a colored bow in the sky. Such a phenomenon is called a *moonbow*.

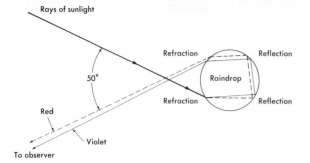

Red

Violet

To observer

Figure 10-6 A secondary bow, of less intensity but larger than the primary bow, is produced as a result of two internal reflections. In the secondary bow reds appear on the inside and blues appear on the outside of the colored arc.

CORONAS AND HALOES

If you develop the habit of watching the sky, you will occasionally see optical features known as coronas and haloes. They occur when a thin layer of cloud covers the sky and partially hides the sun or the moon.

The most commonly observed haloes are in the form of bright rings surrounding the luminary and having a colored or whitish appearance. They occur because of the refraction of light as it passes through ice crystals. For this reason, haloes are an indicator that the cloud layer with which they are associated is a cirrus cloud (Fig. 10-7).

Figure 10-8 shows how a halo is produced. Light from the sun (or the moon) is refracted and reflected by ice crystals. Each of the crystals acts as a six-sided prism. When a widespread layer of ice crystals is present, a large number of particles in a circular area may be in the proper orientation to send the refracted and reflected ray back to the observer. This leads to the ring

Figure 10-7 Cirrostratus cloud producing a halo around the sun. *Photo by* Bob Broder, University of Arizona.

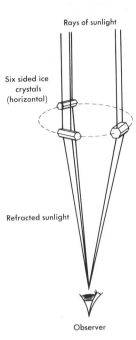

Rays of sunlight

Six sided ice
crystals
(horizontal)

Refracted sunlight

Observer

Figure 10-8   A 22° halo is produced when light is refracted and reflected from ice crystals in the form of hexagonal prisms.

appearance of the halo. The size of the halo is determined by the angle of refraction of the light from the sun or the moon. Most haloes subtend an angular radius of 22°. When a uniform cirrostratus covers the sky, a halo may be in the form of a complete ring. More often the cloud of ice crystals covers only part of the sky, and as a result, only segments of a halo are seen.

As in the prism and the raindrop, the ice crystals act to separate white light into its different colors. Blue waves are bent more than red waves. As a result, the inside of the halo may be red and the outside may be blue. However, since the color separation often is not pronounced, haloes usually have a whitish appearance.

One occasionally sees a species of halo consisting of a faint, white circle passing *through the sun* and running parallel to the horizon. This unusual effect is called a *parhelic circle*. It is caused by simple *reflection* from the faces of ice crystals in the form of hexagonal prisms having their long axes in a nearly horizontal plane.

Sometimes the sun or the moon is partially obscured by a thin cloud and is surrounded closely by a ring of light called a *corona*, a word meaning "crown" (Fig. 10-9). On some occasions the circle of light actually touches the sun or moon, but it is not as brilliant. A corona subtends an angle of only a few degrees and therefore is much smaller than the common 22° halo. Furthermore, when a corona exhibits coloration, it is blue on the inside and red on the outside, just the opposite of the 22° halo.

We have already seen that a halo is caused by a cloud of ice crystals. A corona is produced by a cloud of water droplets. Coronas are not caused by refraction as are rainbows and haloes. Instead, they are a result of a pro-

Figure 10-9 A corona around the sun or the moon occurs because the light is diffracted by water droplets. © Alistair B. Fraser.

cess called *diffraction* by which a beam of light spreads into the region behind an obstacle. The light waves coming from the sun or the moon toward the observer are slightly bent around the droplets. The light waves deviated by many droplets interfere with one another on the observer's side of the droplets and the result is a concentration of light in a circle around the sun or the moon. Whenever you see a corona, you can conclude that the cloud that partly obscures the sun or the moon is made up of water droplets. Roughly speaking, the smaller the cloud droplets, the larger the radius of the corona. This relationship makes it possible to estimate the average droplet radius by measuring the radius of the corona.

When a cloud is made up of ice crystals in the form of plates (Fig. 8-17), they can effectively reflect sunlight and produce unusual optical effects. Some of them occur because crystal platelets tend to fall with their faces in a horizontal orientation. They are like leaves sailing downward along a flat trajectory, perhaps slipping back and forth through the air. Certain clouds of ice crystals resemble a huge ensemble of tiny, horizontally oriented mirrors reflecting the incident rays of the sun. They can produce a luminous streak of light above and below the sun as light is reflected from the bottoms and tops of the crystals. This optical phenomenon is called a *sun pillar* and is most common near sunrise or sunset.

Occasionally you see thin clouds high in the sky that exhibit patches of color of the purest blue, green, and red. They may appear either when the sun is low or high. Such clouds are called *iridescent clouds*. The coloration is usually observed to be within 30° of the sun and is thought to be a diffraction phenomena. The sun's rays are deflected around patches of uniform water droplets, and the light waves interfere with one another in such a fashion as to separate the various color components.

MIRAGES

Desert dwellers have known for a long time that they should not believe everything they see on a hot day. Stories of fiction and reality relate episodes of lost travelers suffering from thirst and being led across the endless sands by the appearance of water that does not exist. The illusion of the presence of water is created by a *mirage*, a refraction phenomenon wherein an image of some object is made to appear displaced from its true position.

There are various types of mirages and all are produced by the bending of light waves as they pass through regions of the atmosphere where the index of refraction of the air changes markedly with height. At times the refraction leads to a distortion of the dimensions of the objects as well as a displacement of their positions.

In discussing rainbows and haloes it was noted that when light rays pass from one medium to another having a different index of refraction, they are refracted. This occurs whenever any electromagnetic wave passes through a region of changing refraction index. It is well known that the index of air, although being close to 1.0003, depends on the frequency of the electromagnetic wave, atmospheric pressure, and temperature. Water vapor is important in the case of radar waves, but it has negligible effects on visible light. On the average, the index of refraction of air decreases with height in the atmosphere. This causes a ray of light coming into the earth's atmosphere, from a star, for example, to be bent downward by a slight amount. When the temperature lapse rate near the ground is unusually small or large, there is unusual refraction of light rays and the occurrence of mirages.

There are two major categories of mirages: *inferior*, and *superior*. The appearance of the mirage depends, respectively, on whether the spurious image appears below, or above, the true position of the observed object. Table 10-3 gives examples of the vertical distributions of temperature and refractive index associated with mirages. Inferior mirages are the most common, usually being discernible whenever the air near the ground is unusually warm and decreases rapidly with height. On summer afternoons this is the normal

Table 10-3 INDICES OF REFRACTION OF AIR NEAR THE GROUND
FOR CONDITIONS FAVORING MIRAGES*

Height (*m*)	Pressure (*mb*)	Inferior mirage		Superior mirage	
		$T\,(^\circ C)$	*n*	$T\,(^\circ C)$	*n*
0	1000	32	1.0002590	22	1.0002678
170	980	24	1.0002607	24	1.0002607
330	960	17	1.0002615	26	1.0002536

*Values of *n* calculated from $n = 1 + 79 \times 10^{-6}(p/T)$, where *p* is in millibars and *T* is in °K.

state of affairs over many highways and deserts. Superadiabatic lapse rates are the rule in the lowest tens of meters of the atmosphere. Figure 10-10 illustrates the paths taken by rays of light reaching the eye of an observer on a day when the temperature decreases markedly with height. A ray of light moving on a downward path in the direction of an observer is refracted upward. As a result, objects such as treetops are displaced downward. Rays of blue light from the sky are bent in such a fashion as to appear as if they are coming off the surface of the earth. They can give the appearance of water on the ground.

Occasionally when conditions for inferior mirages are well developed, the light rays from such tall features as hills and trees that normally are easily seen may disappear from view because the rays of light are deflected over the head of the viewer.

Superior mirages are observed when there is a temperature inversion, that is, the temperature increases with height through a depth of several hundred meters of the atmosphere. In such a circumstance, the refractive index decreases with height and leads to a downward bending of light rays. As a consequence, objects appear above their true positions. Depending on the variation of temperature with height, the images may be inverted and may appear to be smaller, larger, or the same size as the original object (Fig. 10-11).

In extreme cases, the rays of light emitted by an object can be refracted by varying amounts depending on how the temperature varies with height and on the direction of the rays as they leave the object. The result can be a complex combination of inferior and superior mirages that cause

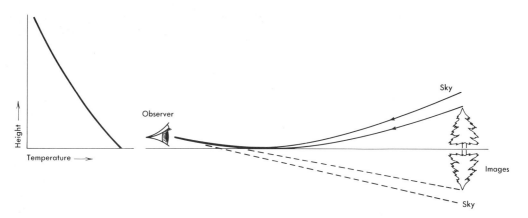

Figure 10-10 When the temperature decreases rapidly with height, an inferior mirage is produced. Light rays from treetops or from the sky follow a concave path and appear to be coming from below the true position of the object (the dashed lines). Blue sky may take on the appearance of water on the ground. In this and the following figure the vertical scales are exaggerated.

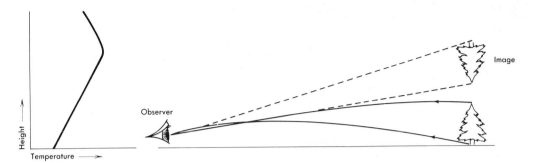

Figure 10-11 When there is a temperature inversion near the ground, a superior mirage may occur. Objects appear higher than their true position. Under certain temperature inversion conditions the images may be inverted (as shown here) and may have their vertical dimensions distorted.

objects such as boats or buildings to appear as towers of great vertical extent. Such a phenomenon is called a *fata morgana*, after the Italian name of King Arthur's story book Sister Morgan le Fay, who could magically build castles out of the air.

LIGHTNING

It is probably safe to say that lightning has been a source of wonder since man first appeared on earth (Fig. 10-12). Imagine the effects on primitive beings when they saw, at close hand, the sudden flash of brilliant light and then heard a loud crash of thunder immediately after the flash. Even today when we know much about the nature of storms and lightning, many people still are frightened by such events. In Chapter 9 it was mentioned that in the United States lightning kills more people, on the average, than either tornadoes or hurricanes. In interpreting these figures it is necessary to keep in mind that every year in this country there are over a million thunderstorms and more than ten times that many lightning flashes. By the simple expediency of getting into a closed shelter such as a building or an automobile, you can protect yourself. Most individuals who are casualties of lightning are those who, during thunderstorms, expose themselves in open spaces or under isolated trees.

There are two general classes of lightning flashes: *cloud-to-ground flashes* that represent a hazard to man and animals and can set fires; *cloud-to-cloud flashes* that often are spectacular to see but do no damage unless they happen to strike an aircraft. Chapter 9 discusses various aspects of thunderstorms and lightning. At this point let us examine some details of the formation of a cloud-to-ground flash.

Figure 10-12 A time exposure showing cloud-to-ground lightning over Tucson, Arizona.

When the concentration of negative electric charge in a cloud and the resulting electrical field strength between cloud and ground are sufficiently large, a lightning stroke is initiated. An avalanche of negative charge surges downward toward the ground. In less than a millionth of a second it advances about 50 m. It then stops for perhaps 50 millionths of a second before advancing another 50 m. In a series of steps it moves toward the surface of the earth; hence it is known as the *stepped leader*.

Fast-responding cameras can record the very rapid progress of the faintly luminous streamer as it moves downward, taking perhaps 0.02 sec to traverse the distance from the cloud base to a height of perhaps 50 m from the ground. Photographs show that the overall luminous diameter of the stepped leader is between 1 m and 10 m, but the electric current flows down a narrower conducting path.

When the advancing tip of the stepped leader approaches the ground, a current of positive electric charge moves up to meet it. Once they make contact, a continuous, electrically conducting channel is produced from cloud to ground. As in the case of any short circuit, there is a sudden, violent movement of electric charges. Negative charges from higher and higher regions of the channel are drained downward to ground. Extremely fast cameras show that the luminous portions of this stroke, known as the *return stroke*, moves upward at tremendous speeds, with the trip from ground to cloud base taking perhaps 0.0001 s.

The peak current in the return stroke often exceeds 20,000 amperes and sometimes exceeds 100,000 amperes. This massive surge of electric charge causes a brilliant flash of light and temperatures that may be as high as 30,000°C. Occasionally there will be only a single stroke such as the one just described. In some cases, a flash may be made up of 30 or 40 strokes and give the appearance of a fast-flickering, electrical arc lasting for perhaps a second. Most commonly a flash has three or four strokes, separated by about 0.04 s with a total flash duration of about 0.2 s. In most strokes the electric current flows for very brief periods, usually less than about 0.0001 s, but sometimes fairly large electric currents continue for periods 10 to 20 times longer. These are the strokes most likely to ignite fires.

In view of the enormous currents flowing in a lightning stroke, it is no wonder that people and animals are electrocuted when struck. Most often the lightning renders its victims unconscious and stops their breathing. When this happens, the immediate use of artificial respiration can often save the life of a person hit by lightning.

Cloud-to-cloud flashes take place between the upper positive and lower negative charge centers of the same cloud or two nearby clouds. It is believed that, initially, a streamer such as the stepped leader bridges the gap between two regions of opposite charge. This is followed by a surge of electrons from the negative to the positive charge region. The discharge channel is luminous for perhaps 0.2 s. During this time there might be several bright pulsations lasting about 0.001 s that correspond to the return strokes in a multistroke cloud-to-ground lightning flash.

Lightning flashes to tall buildings and towers, particularly if they are on mountain tops, as are many radio and television towers, often are started by stepped leaders moving upward from the structure on the ground. The electrical current builds up slowly to a maximum of a few hundred amperes which may last for a few tenths of a second. Occasionally during this period there is a downward surge of current in a return stroke from cloud to ground. At other times no return strokes occur to modulate the relatively long-lasting but low electrical current. Such lightning flashes are distinctive in the sense that, because of the slow buildup of the current, they do not produce the loud clap of thunder usually associated with lightning. The sound level is weak and of a different character than the usual rumble of a thunderstorm. The nature of thunder is discussed later in this chapter.

Over the years observers have noted other occasions of lightning not accompanied by the sound of thunder. You sometimes hear people say that they saw *heat lightning*. It is the luminosity observed from ordinary lightning discharges that are too far away for the thunder to be heard. Such observations often are made on hot summer evenings when the skies overhead are clear. The occurrence of distant diffuse lightning flashes in the apparent absence of thunderstorms may mislead an observer into believing that lightning may be occurring merely as a result of the high temperature. In fact, the presence of the lightning indicates the presence of distant thunderstorms whose formation was favored by the warm, unstable atmosphere. Bright flashes of light at high altitudes can be seen for very great distances, especially if the air is unpolluted. It would not be unusual to see the light from a lightning flash perhaps 50 km away, but thunder is not heard more than about 20 km away and sometimes less than half that distance.

Sheet lightning is not accompanied by thunder when it occurs far from the observer. The phenomenon is seen as a diffuse, sometimes bright, illumination of those parts of a thundercloud within which there has been an intracloud discharge. Sometimes at twilight it is possible to see the grayish appearing tops of cumulonimbus clouds and watch them being illuminated at irregular intervals by interior lightning flashes.

THUNDER

It has long been recognized that lightning is similar to an electrical spark or arc in a laboratory and that it produces a similar type of explosive noise. Only in recent years, however, have we learned about the physical nature of the sound production in a lightning channel.

As noted earlier, in the return stroke there is a great surge of electric charge passing through a channel perhaps 2 cm in diameter. This introduces a tremendous amount of energy and causes the channel temperature to rise suddenly from the cloud temperature which might be about 300°K to perhaps 30,000°K. Within microseconds the pressure increases to as much as 10 to 100 times the value outside the channel. As a consequence, there is a rapid expansion of the heated channel producing a compression wave in the air that propagates outward.

In the first 10 m or so the wave front exhibits a sharp change of pressure and has the characteristics of a *shock wave*, a phenomenon that will be discussed in the next section. After moving away from its source, the pressure wave takes on the form of an ordinary sound wave. Its speed of propagation in the atmosphere depends on the density of the air, that is, mostly on pressure and temperature. At sea level sound travels at a speed of about 340 m/s. This amounts to about 0.21 mi/s (0.34 km/s).

If the speed of sound is known, it is possible to estimate the distance to a lightning stroke. We know that light travels at about 3×10^8 m/s. Therefore, the time taken for the light generated by a discharge to reach an observer is extremely short. For example, if a stroke were 10 km away, one would see the light 33 millionths of a second after it occurred. For most purposes, it can be said that the light is seen at almost the instant it occurs. On the other hand, sound travels very much slower than light, about a million times slower. The thunder from a stroke 10 km distant would take 29 s (10,000 m divided by 340 m/s) to arrive.

Since thunder travels at 0.34 km/s (about 0.21 mi/s), by merely counting seconds between the sensing of light and sound you can easily estimate the distance of a lightning discharge. Every three counts correspond to about a kilometer and every five counts to about a mile. Many people are more frightened by thunder than they are by lightning. The same is true for some cats and dogs and probably other animals as well. Since they cannot be taught to count, their consolation will be up to the owners. People should know, however, that once they hear thunder, the danger from that particular flash is over. Also by counting seconds, you can easily estimate the distance to the lightning. When it is kilometers away and you are inside a suitable structure, there is nothing to fear, providing you are not in contact with or very close to good electrical conductors extending outside the structure. If there is a thunderstorm nearby, stay away from such items as television lead wires or sinks and bathtubs.

When a lightning flash occurs, it emits a wide spectrum of electromagnetic energy of which the visible frequencies constitute only a part. The surges of electric charges along the discharge channel produces substantial power at radio and television frequencies. Such signals are called *sferics*, which is a contraction of an earlier name, atmospherics. When there are thunderstorms in the vicinity, you hear crackling and popping noises on radio receivers. Each audible signal corresponds to a sferic produced by a discharge. If you hear both the sferic and the sound of thunder from the same discharge, you can gauge its distance by counting the seconds between them.

The duration of thunder depends on a number of factors, particularly the location, orientation, and length of the flash channel. Since the return stroke is the source of most of the audible thunder, we should examine how it initiates the shock wave that becomes thunder. Because of the speed of the current through the channel, its expansion starts almost simultaneously at all points in the channel. If it were oriented vertically and you were on the ground, the sound waves from the lowest end of the channel would reach you first because it is closest (Fig. 10-13).

A short segment of the return stroke, perhaps 10 m long, produces a sound wave less than 0.1 s in duration. In the case of a lightning discharge, you can think of each successive 10-m segment contributing to the train of sound waves emanating from the lightning channel. The higher up the channel, the farther from an observer and the longer it takes for the sound wave to reach an observer. In the example shown in Fig. 10-13 the sound from the bottom of the stroke arrives in 12 s (4 km ÷ 0.34 km/s) while that from the top arrives in 15 s (5 km ÷ 0.34 km/s). Therefore, the length of the sound train in the thunder would be about 3 s long. If a lightning channel were very long and tilted away from a listener, the duration of thunder could be several times 3 s. In addition, sound reflections from the terrain could extend the rumbling sound for many more seconds.

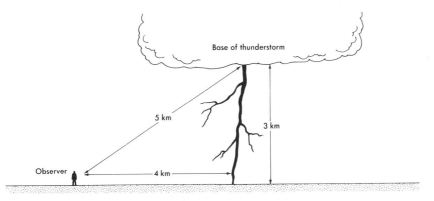

Figure 10-13 Sound waves generated by the upper part of the lightning stroke reach the observer later than the ones generated by the bottom of the flash.

SONIC BOOM

The advent of the supersonic airplane introduced the *sonic boom*, an explosive sound produced by aircraft flying at speeds exceeding the speed of sound. An airplane traveling at slower speeds produces noise that propagates in all directions faster than the forward speed of the airplane. As a result, there is a continuous noise of the kind associated with a typical subsonic airplane flying overhead.

When an object moves at a speed exceeding that of a sound wave, the object is constantly advancing faster than is the pressure wave it is producing. In a sense, a supersonic airplane is forcing the air ahead of it to move abruptly without earlier movement. This causes a narrow zone of compressed air in the form of a shock wave through which the pressure changes abruptly over a small distance. As illustrated in Fig. 10-14, the forward edges of an airplane produce a bow wave with a sudden rise in pressure. A tail wave characterized also by a pronounced pressure increase is associated with the trailing surfaces of the airplane. The sudden pressure changes are indicative

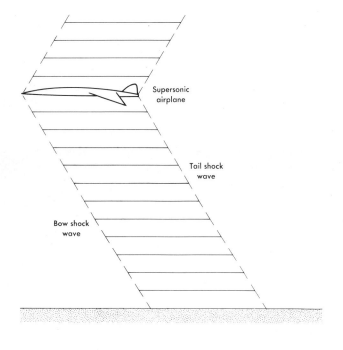

Figure 10-14 When an airplane flies faster than the speed of sound, it creates shock waves across which there are abrupt pressure changes. They account for sonic booms.

of shock waves and each of them produces a sonic boom in much the same way that an explosion does.

When an airplane flies along at a supersonic speed, it leaves below and behind it a wave of sonic booms. A large commercial supersonic airplane (called an SST) can produce sonic booms along a track perhaps 80 km wide with the pressure change and boom intensity being greatest along a track under the airplane.

The various surfaces of an SST set up individual shock waves that can be discerned close to the airplane, but when such an airplane is at usual flight altitudes of perhaps 20 km, only the bow and tail sonic booms are distinguished. They travel along at the same speed as the airplane.

The boom intensity depends on the peak over-pressure, and it in turn depends on the speed, size, and flight altitude of an airplane. A supersonic airplane might produce a peak overpressure of about 0.5 mb at the ground. It is a small value in absolute measure, but such a pressure wave produces a noise level about a million times greater than that of ordinary conversation at a distance of 1 m. Most of the sound energy in a sonic boom occurs at about 10 hertz, frequencies below the lower limit of human hearing which is about 20 hertz. On the other hand, the low-frequency pressure fluctuations cause vibrations of windows, doors, and walls. Occasionally a particularly strong sonic boom may even cause damage to buildings, particularly the cracking of plaster and glass.

As long as an airplane flies at speeds faster than sound, it will produce a sonic boom at the ground. This is one of the reasons some environmentalists are opposed to the use of supersonic transports. If an airplane is well designed and if the flight path is selected carefully, the effects of sonic booms can be reduced but they cannot be eliminated. In 1976 the Soviet TU-146's and English–French Concorde SST's were introduced into commercial service. In addition to sonic booms, there is concern about the noise they produce on takeoffs and landings as well as about the effects of engine emissions on the ozone concentrations in the upper atmosphere (see Chapter 2). Time will tell if the advantages of moving around the world at twice the speed of sound are large enough to overcome the problems of sound, gaseous, and particulate emissions.

Unlike thunder, sonic booms are not produced by processes in the atmosphere; they come into being because of the compressibility and the fluid properties of air. Next time you hear an explosive sound on a day when there is no lightning in the vicinity, look to the sky. The source is likely to have been an airplane flying at supersonic speeds. You probably will not see it because it will be at a great altitude, but the boom makes its presence likely.

CHAPTER 11

Climates of the Earth

WHEN A METEOROLOGIST TALKS ABOUT THE WEATHER, he refers to short-term variations of the state of the atmosphere. This would include consideration of air temperature, humidity, cloudiness, precipitation, winds, and other atmospheric properties as they change from minute to minute or even from one month to the next.

The long-term manifestations of the weather are considered to be part of the climate. Stated in another way, the climate of any region is represented by the statistical properties of the weather over an extended period, typically several decades. In order to understand the climate of a particular place, you have to know more than just the average values of temperature and rainfall, even though these quantities are most informative. It is necessary to know the averages of such relevant meteorological factors as humidity and wind velocity as well as the variations of all important weather elements from one period to the next.

Climate is in a constant state of change. As will be seen later in this chapter, paleoclimatologists have shown that the climate of our planet has undergone major alterations over its several billion year history.

When one speaks of the climate, it is essential to delineate the region involved. In recent years a great deal of attention has been given to the study of the climate of the entire earth. More traditionally, climatologists have concerned themselves with the climate of states, countries, or parts of continents. On a smaller scale, studies have been made of the climate inside a single building, such as a house or a barn, or over a field of grain.

SCALES OF CLIMATE

The term *microclimate* is often used to represent the climate of the atmosphere between the earth's surface and a height where the earth's influence becomes indistinguishable. Such a layer, measured in terms of tens or hundreds of meters, is of great importance in many practical problems. This is the region where people live and vegetation grows. The number of bushels of corn or wheat yielded by a field is determined to a significant extent by the microclimate of the field.

Figure 11-1   Standard shelter for weather instruments. *Courtesy* Science Associates, Inc.

The time of occurrence of frosts in the spring and in the fall are crucial climatological events to agriculturalists. The length of the growing season is determined by the period of time between the last frost-free day in the spring and the first occurrence in autumn of a frost of sufficient intensity to damage the plants. The temperatures at which this occurs depends on the species and the stage of development of the vegetation.

When data on killing frosts are being evaluated, it must be recognized that the temperature of the leaves can be significantly different from the official temperature reported by the National Weather Service. These values are obtained by means of thermometers, about a meter and a half above the ground, mounted in an instrument shelter (Fig. 11-1).

During the sunny part of the day the leaf surfaces can have temperatures substantially higher than the official temperature reading. As a result, the plants transpire actively. The stomates on the leaves open and water vapor is lost to the air. This serves as a mechanism for cooling the plant. At the same time, some of the incident solar energy, not more than about 2 percent, is used in the process of photosynthesis by which the plant takes up carbon dioxide and grows new leaf cells. In order for vegetation to thrive on hot, sunny days, especially if the air is dry, there must be adequate soil moisture to move through the root system and up the plant to the leaves.

At night, especially when the air is dry and the sky is clear, temperatures near the ground and particularly the leaf temperatures can be substantially lower than the air temperature recorded in an instrument shelter. As noted in Chapter 5, outgoing infrared radiation from surfaces exposed to the night sky can lead to pronounced low-level temperature inversions. The temperatures of the radiating surfaces can be several degrees lower than the air just a few meters above them. This accounts for the occurrence of dew and frosts on nights when the official weather observations would indicate that they should not have happened. In desert areas it is not unusual to observe frosts on nights when the official minimum temperature is +1°C (34°F) or so.

Agriculturalists recognize these facts and have developed techniques that can combat certain frosts. On nights when the winds are light and outgoing infrared radiation is acting to lower plant temperatures to dangerously low levels, there are several methods that can be used. Fruit orchards can be lined with heaters (Fig. 11-2) that warm the air and the trees. Some citrus growers in the United States use power-driven propellers mounted on masts perhaps 5 m high to stir the air in the temperature inversion layer and bring the warm air down to the tree level. Certain low-growing berries can be saved from killing frosts by spraying them with water or flooding the fields continuously through the night. If temperatures do not go too low, the plant can be kept near 0°C and preserved from a killing freeze. If the air temperature falls into the low twenties Fahrenheit (for example, −5°C) and there is a moderate or strong wind, little can be done to protect temperature-sensitive fruit trees. The degree of damage depends on the type of vegetation and the stage of growth.

To an increasing extent, glass and plastic greenhouses are being used for growing vegetables and flowers. Since the interior climates can be controlled to a large extent, greenhouse farming can yield, for each acre under cover, very large quantities of produce such as tomatoes and cucumbers.

Figure 11-2 Heaters in a citrus orchard in Arizona.

The climatic characteristics of regions perhaps 10 km to 100 km in size can be classified as part of a *mesoclimate*. For example, the atmospheric properties of a valley or of a city fall into this category. It has long been known that the climate of a city differs in a number of important respects from the climate of the nearby countryside. Traditionally, particularly in the days before air conditioning became available, many people who wanted to escape the heat of summer and who could afford the costs had second homes away from the cities.

During recent years meteorologists have been studying in great detail how cities affect the local climate. One approach has been to observe atmospheric conditions in and around a large industrialized urban area such as St. Louis, Missouri. It has been found that, as is the case with most large cities, St. Louis has a marked *heat island*, that is, air temperatures through a layer perhaps 1 km deep average about 1°C warmer than the air temperature in surrounding regions. At the same time, the relative humidities are lower than in the adjacent rural areas. The primary reasons for these temperature and mositure anomalies are the destruction of vegetation and the paving of streets, sidewalks, and parking lots and the construction of buildings.

The cement and asphalt absorb heat during the day and radiate it slowly at night, but much of it is not directly radiated to the night sky. The release of heat into the atmosphere by most human activities also contributes to the warmth of the city, but it is considered a secondary factor compared to the storage of absorbed solar energy in the urban structures and pavements.

Water runs off asphalt and cement faster than it does off soil and vegetative covers. As a result, within a city there is less water for evaporation and transpiration and humidities of urban air are reduced. Since less of the available heat is used for evaporation and transpiration, the heat serves to increase the temperature of the city and the air over it. This factor contributes importantly to the heat-island effect.

Interestingly, during the summer, rainfall over and downwind of large industrialized areas such as St. Louis is greater than would be expected if the city did not exist. Several reasons are given for this. Increased temperatures stimulate convection. The frictional effects of the city on air moving over it also lead to low-level convergence and augmented cloud formation on days when the atmosphere is naturally unstable and humid.

A fascinating study of the effects of a city on the local climate has been conducted under the direction of the outstanding American climatologist, Helmut E. Landsberg. He and his associates at the University of Maryland have been measuring the meteorological properties of the air over Columbia, Maryland. They started the measurements when the town was being planned and have watched them change as the community has grown. Figure 11-3 shows how the temperature increased over the 7-year period starting in 1967. At that time there were only 200 inhabitants. By 1974 there were about

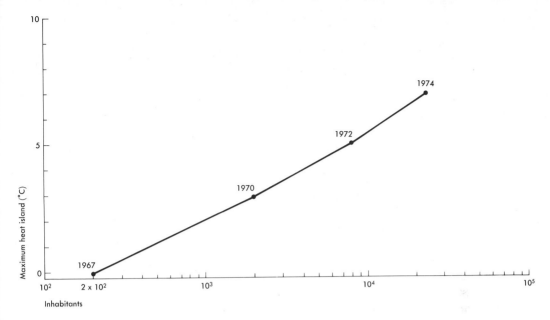

Figure 11-3 The maximum value of the heat island, i.e., the difference between urban and rural temperatures, on calm, clear evenings at Columbia, Maryland, correlated to population. *Courtesy* H. E. Landsberg, University of Maryland.

24,000 residents, and the maximum value of the heat-island effect had increased to 7°C and was still rising. In 1976 it was estimated that by 1980 the population would reach about 100,000 and the heat island would amount to 10°C.

The term *macroclimate* can be used to describe conditions over large areas, perhaps covering a state, country, or even a continent. When considering the earth as a whole, it is common to speak of the global climate. Many scientists have been studying the causes of the overall warming and cooling periods that have occurred in the distant and recent past and are occurring today. This subject is addressed later in this chapter.

TEMPERATURE

Traditionally, climatology has involved the analysis of large amounts of data. Temperature and rainfall measurements from stations all over the world have been accumulating for a long time. In some places they go back some two centuries or more, while in others the record is measured only in decades or less. Although there are some island stations having long records and in some places there have been shipboard measurements, for the most part observations over the oceans are inadequate for the construction of adequate climatologies.

The climate of a region is described, in part, by the averages of meteorological quantities over periods on the order of 30 years. As noted earlier, a complete description of the climate should also include information on the variations of these quantities throughout the year as well as over longer time periods. The importance of knowing these variations can be illustrated by the curves in Fig. 11-4 which show the mean monthly temperatures at San Francisco, St. Louis, and New York City. The annual average temperatures of these cities are close to one another, but the climates are different. St. Louis

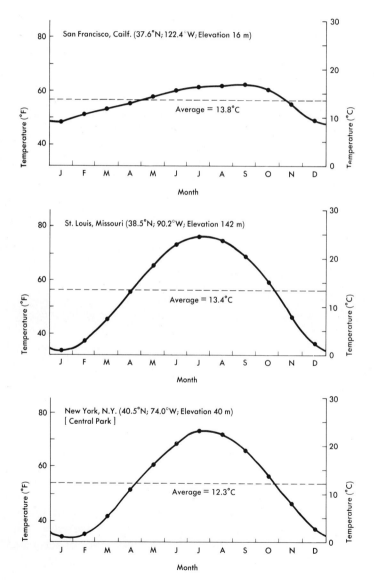

Figure 11-4 Average monthly temperatures in San Francisco, St. Louis, and New York City. The dashed line shows the annual average temperature.

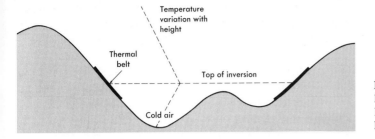

Figure 11-5 Cold air sinks into the lowest places. Temperatures are higher in the thermal belts than at lower elevations.

and New York have more extreme temperatures; they are hotter in summer and colder in the winter than the west coast city. Appendix III gives average monthly temperatures at cities other than the ones shown in Fig. 11-4.

The differences in temperature from one place to the next often, but not always, can be explained in terms of differences in one or more of the following factors: altitude, latitude, and proximity to a large body of water.

In general, the higher the station, the lower the temperature. This obviously comes about because in the atmosphere the temperature usually decreases with height. In places where surface temperature inversions are common, such as in desert regions, the normal relationship of altitude and temperature does not hold for the lowest few hundred meters of the atmosphere. The average temperatures at stations along the slopes of mountains or on hilltops are often higher than in the valley bottoms. At night the coldest, heaviest air drains toward the lowest regions and the slightly higher elevations are exposed to the warm air near the upper parts of an inversion layer (Fig. 11-5). These warmer regions are called *thermal belts*. Citrus orchards are more likely to survive the effects of radiative cooling when they are located in thermal belts than if they are in the valley floors. This certainly is the case in the arid regions of the southwestern United States and other similar climatological regions.

The relationship between a station's temperature and latitude is explained, to a large extent, in terms of the position of the sun. As noted in Chapter 3, annual amounts of insolation and the average radiation balances of the earth's surface are highest in equatorial regions and decrease toward the poles. As a consequence, average ground-level air temperatures are highest at low latitudes and tend to decrease toward the poles. The configurations of continents and oceans and elevation differences have important effects on the latitudinal gradient of temperature. Furthermore, the characteristics of the general circulation significantly influence the global temperature patterns because air currents generally transport heat poleward (Fig. 3-10).

A comparison of the average sea-level air temperature in January and July (Fig. 3-12) shows that the seasonal range of temperature (the difference between summer and winter temperature) also varies with latitude. The range tends to be small in the equatorial zones and increases with increasing latitude. This occurs mostly because the month-to-month variation of incoming solar radiation is small at low latitudes and increases with increasing latitude.

As shown by the temperature data in Fig. 11-4 and in Appendix III, maximum annual temperatures in the Northern Hemisphere are usually found in July even though maximum incoming solar radiation occurs in June. In most places minimum temperatures are observed in January; they lag the solar radiation minimum by a month or so. These results are explained by recognizing that as long as incoming heat flux exceeds outgoing heat flux, the temperature of a volume of air continues to increase. Certain authors have sought to show that the observed variations of temperature from month-to-month are a result of differences of incoming solar radiation and outgoing terrestrial radiation at the Earth's surface. Over tropical and subtropical regions, however, insolation exceeds outgoing terrestrial radiation every month of the year. Quite evidently the annual march of temperature cannot be explained in the terms of radiation alone.

In order to account in a quantitative sense for temperature changes through the year, it is necessary to consider all significant sinks and sources of heat and other mechanisms of heat transfer in addition to radiation. As was noted in Chapter 3, incoming solar radiation is mostly absorbed by the ground. In the spring and summer the soil and water at the Earth's surface are warmed to increasing depths. They become sources of heat. It is transferred into the atmosphere by means of infrared radiation as well as by conduction and convection. The temperature at any place also depends, to an important extent, on heat transfer by air and ocean currents.

The same comments apply to *diurnal*, that is daily, variations in temperature. These obviously depend to a large extent on incoming solar radiation, but maximum temperatures generally occur several hours after noon, the time of maximum insolation. Minimum temperatures are usually experienced 30 min to an hour after sunrise. In accounting for hour-to-hour temperature changes, heat storage by soil and water bodies and heat transfer by means of conduction and convection are important. Diurnal temperature variations are greatly influenced by the location of a station in relation to large bodies of water. In cloudy, wet regions near oceans the diurnal temperature range is small. It is large over low-latitude, desert areas.

The temperature maps in Fig. 3-12 show that the proximity of oceans has a profound effect on a station's temperature. For example, see how air temperatures along the coasts are influenced by the temperatures of the coastal waters. As was noted in Chapter 3, the oceans are tremendous heat reservoirs. Their temperatures change less than do the temperatures of the continents. As air passes over the oceans, its temperature is modified and approaches the water temperature. Localities immediately downwind of the oceans, such as those along the west coasts of continents, are largely controlled by the ocean temperatures. This accounts for the fact that the annual range of air temperature at San Francisco (Fig. 11-4) is small. It is said to have a *marine climate* because of the major influence of the sea.

On the other hand, a station such as St. Louis, which is far from the oceans, has a large annual range of temperature. Figure 3-12 shows that places at the interior of large continents can have extremely large temperature changes from winter to summer. They are said to have *continental climates*. Outside of Antarctica the lowest recorded surface air temperature is the −67°C measured in highly continental Verkhoyansk, Siberia. It also has recorded summer temperatures as high as 32°C.

Figure 11-6 shows the temperature profiles of London, Moscow, and Tokyo, further illustrating the points made in relation to the three cities in the United States. The maritime nature of London's climate should be expected because England is an island. Note, however, that since London is not immediately downwind of the ocean, it has a greater range of temperature than does San Francisco. London is about 13.7° of latitude (about 1500 km) farther north and therefore it is somewhat cooler on the average than San Francisco. Moscow, far from the oceans, experiences a large range of temperature from winter to summer as does any station having a continental climate.

The annual march of temperature at Tokyo is similar to the one at New York City, shown in Fig. 11-4. This might not be surprising considering the fact that both are located along the eastern extremities of large continents. Nevertheless, since Tokyo is an island, one might have expected its temperature range to be smaller than the one at New York. In the winter the massive Asian continent is the source of strong, persistent westerly winds bringing large bodies of cold air that cannot be warmed a great deal, for they pass over the Sea of Japan that separates Japan from the mainland.

Average summer temperatures in Tokyo are relatively high. This is, in part, a consequence of the warm water in the northward flowing Kuroshio Current (see Fig. 3-14). If it did not exist, temperatures in winter would be even colder than they are now. Similarly, the Gulf Stream, off the east coast of North America, serves to raise the average annual temperature of regions near the sea. In the Southern Hemisphere the Brazil Current and the Agui Current warm parts of eastern South America and eastern Africa, respectively.

On the other hand, cold, equatorward currents act to reduce summer temperatures of nearby locations. The southward moving cold California Current off the western United States accounts, in part, for the relatively cool weather of San Francisco and other places along the west coast of the United States. The low temperature of the ocean water is partly a result of *upwelling*, the rising of cold subsurface water to the surface. This occurs because of persistent wind stresses on the surface water carrying it away from the coast. Upwelling also is common off the west coast of Africa and South America.

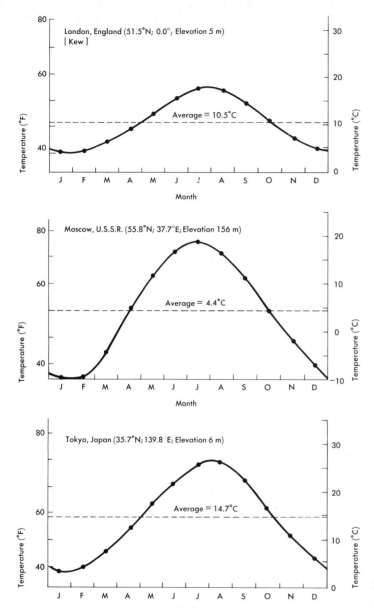

Figure 11-6 Average monthly temperatures in London, Moscow, and Tokyo.

Upwelling brings up cold, nutrient-rich water from the lower depths of the ocean. This accounts for the fact that one of the most productive fishing regions of the world is located off the coast of Peru, Ecuador, and Chile. The wind currents over this area periodically deviate from the normal, leading to changes of ocean currents and a suppression of the upwelling. Warm water moves into the region from the north. This condition, known as *El Niño*, usually occurs in February and March and is highly variable in

Figure 11-7 A low fog partly obscures the Golden Gate Bridge in San Francisco, CA. *Courtesy* L. Blodget, San Francisco Convention and Visitors Bureau.

extent and duration. When there is an El Niño, the reduction of nutrients leads to a reduction in the fish population. In extreme conditions there can be drastic losses to the fisheries and other activities that depend on an abundant fish supply.

Incidentally, along the California coast the cold water accounts for the frequent occurrence of fogs and low stratus clouds. They are particularly common in late spring and early summer. Warm, humid air passing over the ocean is cooled and its relative humidity rises. In the late afternoon and through the night there is radiation cooling to augment the ocean's effects. As a consequence, atmospheric humidity approaches saturation and condensation occurs. Often the fog and stratus drift over the land from the sea and remain until the warming caused by solar radiation is enough to evaporate the droplets (Fig. 11-7). The process can repeat itself for days on end, with the fog rolling over the land in the late afternoon and "burning off" in the morning.

RAINFALL

In Chapter 8 a discussion was given of the physics of rain, snow, and hail and the movement of water substance through the geophysical system of land, ocean, and atmosphere. This subject has been given a great deal of attention by climatologists, meteorologists, and hydrologists for a long time. They have developed a large body of knowledge about the climatological characteristics of precipitation.

Figure 11-8 Standard stick-type rain gauge used by the U.S. National Weather Service. *Courtesy* Science Associates, Inc.

As every farmer knows, it is easy to measure rainfall. An ordinary bucket or tin can placed in the open, away from trees or buildings, can give reliable measurements. The instrument used by the National Weather Service over the years is, in a sense, a glorified tin can. (Fig. 11-8). It has an opening 20 cm in diameter that funnels the water into a cylinder whose cross-sectional area is one-tenth that of the gauge mouth. As a result, a 1-cm rainfall will produce a column of water 10 cm deep in the cylinder. An appropriately calibrated stick is used to measure the depth of water accumulated in the gauge and obtain the quantity of rainfall.

Figure 11-9 shows one of a variety of recording rain gauges. Precipitation falling into the top of the instrument is funneled to a pail sitting on a scale. It weighs the water as it accumulates. As the scale is depressed, it

Figure 11-9 A recording rain gauge; rain or snow falls into a bucket and is weighed. See Fig. 11-10. *Courtesy* Science Associates, Inc.

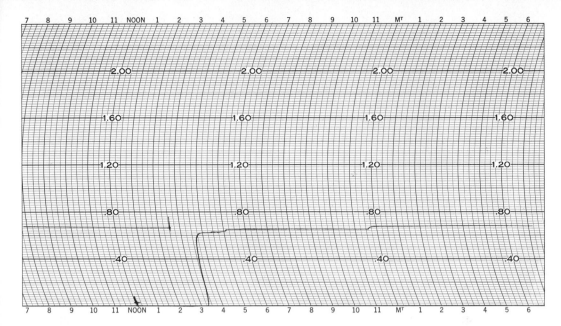

Figure 11-10 Record made by a weighing gauge such as the one in Fig. 11-9. The chart covers a period of about 24 hours beginning 12:05 P.M. July 18, 1963. Note that between about 3:15 P.M. and 3:45 P.M. there was 0.60 in. (15.2 mm) of rain. Two other brief, light showers occurred at 5:05 P.M. and 11:35 P.M.

moves a pen across a chart mounted on a rotating drum that is turned by a clock (Fig. 11-10). By means of a recording gauge, it is possible to measure the rates of rainfall over short periods of time. For some purposes it is necessary to know not only the total rainfall but also the time over which it falls. This is particularly true in the case of thunderstorms over hilly or mountainous terrain. In such circumstances, flash floods may come into being quickly—in an hour or two—and turn a quiet stream or river into a dangerous torrent.

The amounts of rain that can fall in extreme cases stagger the imagination. Table 11-1 gives some statistics for the maximum accumulations for various periods of time. As you would expect, the heaviest rain intensities are produced by cumulonimbus clouds in warm, humid tropical air masses where the tropopause is high. This allows thunderstorms to grow to great altitudes, and in the process large masses of water are condensed. Great depths of rainfall *at one place* can occur when large, long-lasting storms stagnate or move slowly. The last three entries in the table indicate that rainfall is greatest over mountains in tropical regions. Cumulonimbus clouds persist over the high terrain and produce steady streams of heavy rain.

The quantity of snow is expressed in several ways. One common procedure is to specify the depth of snow accumulated over a given period of time. More commonly, the snowfall is expressed in terms of the depth of water resulting when the snow is melted. For hydrologic purposes, this is the more satisfactory of these two specifications of snowfall, because it indicates the mass of water that has reached the ground.

If you have lived in regions where snow is not a rare event, you know that the properties of snow may vary a great deal from one storm to the next. Sometimes when temperatures are very low, snow falls in the form of dry, individual ice crystals. You can catch them on a piece of cloth and examine their intricate hexagonal structures. On other cold occasions snowflakes are aggregates of dry snowflakes made up of many ice crystals that have collided and adhered.

Table 11-1　SOME EXTREME RAINFALLS

Time period	Rainfall		Location
	(cm)	*(in.)*	
1 min	3.1	1.2	Unionville, Maryland
15 min	20	8	Plumb Point, Jamaica
42 min	30	12	Holt, Missouri
4 hours	53	21	Basseterre, St. Kitts, W.I.
12 hours	135	53	Belouve, La Réunion Island
24 hours	188	74	Cilaos, La Réunion Island
1 month	930	366	Cherrapunji, India
1 year (average)	1168	460	Mt. Waialeale, Hawaii
1 year (maximum)	2647	1042	Cherrapungi, India

Sources: Technical Report 70-45-ES, Earth Sciences Laboratory, U.S. Army Engineers, and Hydrometeorological Report No. 5, U.S. Weather Bureau.

Sometimes when the air temperature at the ground is just a few degrees less than 0°C, snowflakes are relatively heavy, somewhat moist aggregates. On such days snowballs pack readily and snowmen are easy to make. In these cases, the density of freshly fallen snow is greater than it is when the snow is dry and light. "Wet snow" may have densities of about 0.15 g/cm³. Since the density of water is about 1 g/cm³ (Table 2-4), a 7-cm depth of this snow is equivalent to a rainfall of 1 cm. On very cold days the snow is dry and its density would be less than 0.1 g/cm³. As a rule of thumb, 10 cm of snow can be taken to be equivalent to 1 cm of rain.

By melting the snow and measuring the depth of water, meteorologists circumvent concerns about snow density and obtain direct measurements of the volume of water. This is what hydrologists and water engineers need to know as they estimate quantities of water stored in frozen form over mountain watersheds.

As would be expected, snowfalls are greatest in regions where temperatures fall below 0°C during the winter and where humid air passes over mountain ranges. As the air rises, clouds form and ice crystals grow, collide with one another, and fall as snowflakes. In some circumstances very large quantities may accumulate. During one record-setting 24-hour period,

Silver Lake, Colorado, experienced 193 cm (76 in.). During a single winter season, Paradise Ranger Station in Washington measured 2608 cm (1027 in.). Note that if you use the one-tenth rule cited above, these quantities are still substantially smaller than the extreme rainfalls for the same periods given in Table 11.1.

A crucially important aspect of precipitation is its tremendous variability from place to place and from time to time. The distribution of average annual precipitation over the earth is shown in Fig. 11-11. It ranges from well below 25 cm to much more than 250 cm. The smallest values are found in the desert areas under the eastern parts of the semipermanent, subtropical anticyclones (see Fig. 6-2). The prevailing subsidence inhibits clouds and rain. Some prominent examples of extreme dryness are given in Table 11.2. A nearly incredible example of aridity is the experience of Iquique, Chile, which reportedly had no rain for 14 consecutive years.

Table 11-2 AVERAGE ANNUAL PRECIPITATION IN SOME EXTREMELY ARID PLACES

	Precipitation	
Places	*(cm)*	*(in.)*
Arica, Chile	0.08	0.03
Wade Halfa, Sudan	Below 0.25	Below 0.1
Bataque, Mexico	3.0	1.2
Aden, Aden	4.6	1.8

Source: Technical Report 70-45-ES, Earth Sciences Laboratory, U.S. Army Engineers.

The heaviest rainfall amounts shown in Fig. 11-11 occur in mountainous, equatorial, and tropical regions where showers and thunderstorms frequently occur in unstable, warm, humid air. In general, the higher the station, the greater the annual precipitation. Hills and mountains may serve as barriers causing the air to rise and clouds and rain to form. During summer days air temperatures over sun-warmed highlands are higher than at the same elevation over the adjacent valleys. As a result, convective clouds, showers, and thunderstorms are more frequent over the higher terrain. In the case of very high mountains extending to altitudes where atmospheric humidities are low, precipitation on the summits may be less than at intermediate elevations.

The annual distribution of precipitation can differ greatly from one place to the next. Figure 11-12 gives several examples and Appendix III gives many more. Most of the precipitation in San Francisco falls from cyclonic storms moving in from the Pacific during the winter. The summer months are extremely dry. In some localities, such as St. Louis, Missouri, precipita-

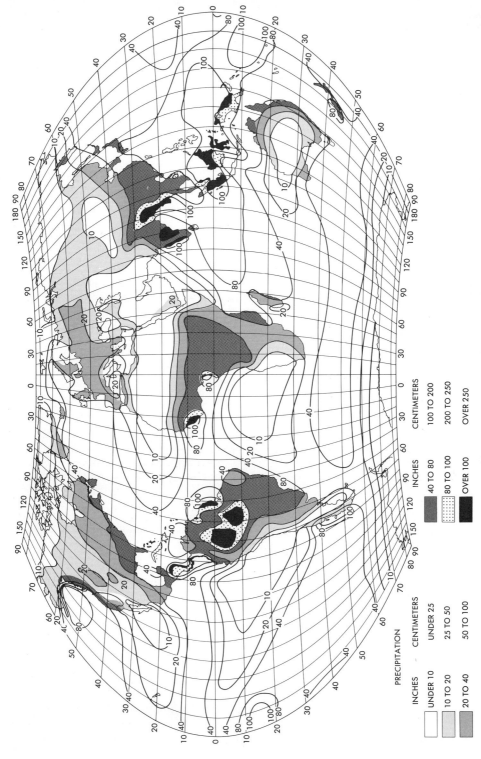

Figure 11-11 General pattern of average annual precipitation over the earth. Isohyets are labeled in inches. See conversions to centimeters. From *Climates of the World*, U.S. Government Printing Office, 1972.

PRECIPITATION	
INCHES	CENTIMETERS
UNDER 10	UNDER 25
10 TO 20	25 TO 50
20 TO 40	50 TO 100

INCHES	CENTIMETERS
40 TO 80	100 TO 200
80 TO 100	200 TO 250
OVER 100	OVER 250

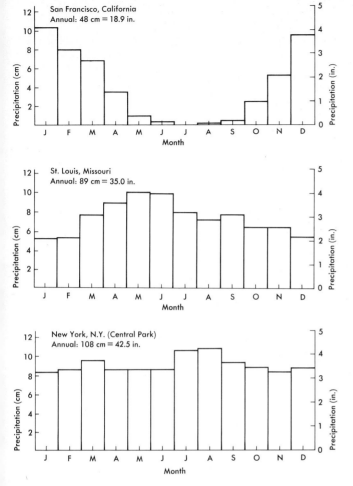

Figure 11-12 Average monthly precipitation at San Francisco, St. Louis, and New York City.

tion is a maximum in the spring and early summer, mostly from showers and thunderstorms. In still others, such as New York City, precipitation normally falls throughout the year with relatively little change from season to season.

The character of the annual variation of rainfall is an important factor in the classification of the climate of any region. More importantly, it can exert vital controls over the farming activities of the region. In arid regions the scarce and highly variable rainfall makes it necessary to use irrigation in order to grow cotton, lettuce, melons, and other agricultural products. The water is in short supply and is best used for the production of relatively high-priced crops. On the other hand, in areas such as the midwestern United States the land is relatively flat and fertile and usually receives a plentiful supply of precipitation during the growing season. St. Louis, Missouri, in the heart of corn and soybean country gets an average of about 89 cm per year, much of it falling during the growing season.

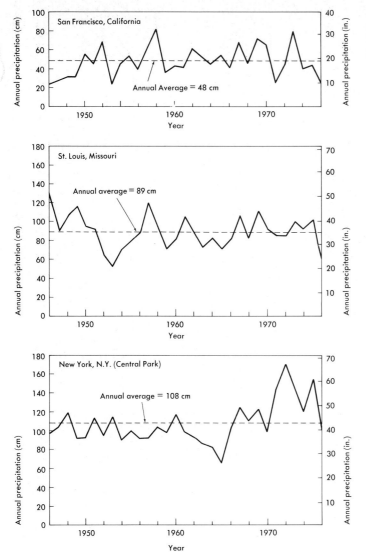

Figure 11-13 Annual precipitation amounts at San Francisco, St. Louis, and New York City for the years 1946 to 1976.

Unfortunately, as everyone knows, the *average* annual amounts of precipitation seldom occur. This fact is clearly shown in Fig. 11-13 which depicts annual precipitation at various places over about three decades. The large variations from year to year are typical and occur at most stations all over the world. These data clearly demonstrate that the so-called *normal annual precipitation* is an unusual event because in any one year precipitation is usually greater or less than the *normal*.

DROUGHTS AND FLOODS

As long as rainfalls are close to the average, most economic activities—agriculture, industry, etc.—can function effectively, because, in general, society adapts to the prevailing conditions. But when precipitation amounts deviate substantially from the average, there can be serious problems. This is especially true when precipitation is below the average for a prolonged period over a large region. Such an event is called a *drought*. This century has seen many droughts, some having major consequences on the lives of millions of people. During the 1930's the United States suffered the worst drought of its history (Fig. 11-14). In those years rainfall over the central parts of the country was substantially below normal. The soil was dry; the plants died and loosened the anchors holding the ground together. Strong winds scoured the earth, layer by layer. Huge clouds of pulverized topsoil were transported over thousands of kilometers. This was the time of the great Dust Bowl.

Since the 1930's there have been periodic droughts over various regions of the United States. For example, in the 1960's the northeastern United States suffered from inadequate precipitation; in the late 1970's the western United Stetes suffered a serious drought. Other countries have not been spared. In 1972 and 1975 the Soviet Union suffered major damage to its grain production because of inadequate rains. The Sahel region of Africa was victim to extreme drought in the early 1970's at a cost of misery and

Figure 11-14 One effect of a drought. *Courtesy* National Oceanic and Atmospheric Administration.

starvation to a great many people. Other countries such as India saw years of dryness and a lack of adequate food in the middle 1970's.

The term drought is defined in the dictionary as "prolonged dry weather." It is reasonable to ask, "How prolonged?" and "How dry?" It just is not possible to give an all-inclusive answer, for the answer depends on how the rain is used. If corn, wheat, or some other crop dies in the field for lack of water, a drought has occurred. If cattle on an open range suffer because of empty water holes, the same can be said. When the city of New York has to restrict lawn sprinkling and the operation of air conditioners, the restrictions can be blamed on a drought over the city's watersheds.

As a general observation, it is fair to say that when living organisms are suffering because of a deficiency of rainfall, a drought is occurring. Thus, the criteria for the presence of a drought are more biological than meteorological, even though weather factors bring it about. In some regions, such as the northeastern United States, 50 cm (20 in.) of rain in a year represents drought conditions. The same quantity of rainfall in southern California would represent a year of water abundance. The damage suffered by plants during a drought depends not only on the rainfall but also on the species of the plant and the properties of the soil. Therefore, even in a single region a prolonged dry spell may harm some plant communities and not others.

Meteorologists have sought to derive definitions of a drought based on rainfall. No definition has been entirely satisfactory because of the complicated and variable relationship of weather, plants, and animals. But, obviously, the smaller the actual rainfall in comparison to the normal rainfall, the greater the likelihood of a drought and the more serious its consequences.

Since rainfall deficits are the basic cause of droughts, some procedures define a drought according to the length of time and degree of abnormally low precipitation. For example, a drought might be simply defined to exist when the annual precipitation is less than 80 percent of normal for two or more successive years. On this basis, St. Louis and the surrounding area suffered a serious drought in the early 1950's. In fact, this did occur with great losses to agriculture. The northeastern United States experienced a major drought in the 1960's, as indicated by the data for New York City in Fig. 11-13.

A drought index widely used by agriculturalists was devised by Wayne C. Palmer, an American meteorologist. It takes into account the quantity of monthly rainfall needed to maintain normal quantities of soil moisture in any particular locality.

An important point to recognize is that periods of rainfall deficits and droughts are a normal feature of a meteorological history at any place. They have occurred in the past and will occur in the future, but it still is not possible to predict accurately when they will occur. Prudence advises that in planning for the water and food supplies of any region or for the world as a whole the inevitability of droughts must be recognized. Surpluses during years of good weather should be stored for use during years of shortages.

Droughts occur when the character of the atmospheric circulation is abnormal and, as a result, regions where rising air is regularly expected are dominated by sinking air that inhibits cloud and precipitation formation. On weather maps the likelihood of abnormal patterns of rain and snow is evidenced by shifts in the patterns of pressure troughs and ridges. The troughs, where pressure is low, are usually associated with wet areas. If they are supplanted by persistent pressure ridges and anticyclones, a drought is in the making. Reasons for major, persistent shifts in the general circulation of the atmosphere still are being debated. They are of critical importance because they determine whether temperatures and rainfall are nearly equal to normal or if there are likely to be droughts or floods, the almost equally devastating events at the opposite end of the hydrological spectrum.

There are various scales of flooding, both in time and space. Heavy rainfalls from thunderstorms can fill stream channels and cause flash floods. Although they sometimes can be devastating as was the case in Rapid City, South Dakota, on June 9, 1972, flash floods are usually fairly local in nature and affect a relatively small area and just a few people. When deluges occur over the drainage basins of the large river systems, water can overrun the banks and spread over vast regions. The Ohio, Missouri, and Mississippi river systems experience such flooding almost regularly.

The general sequence of events leading to major floods in the United States has been repeated time and again. During a cold, wet winter a blanket of snow is deposited over the Middle West. Temperatures are low, the ground freezes, and snow accumulates with relatively little melting. In the early part of the year a warm, humid mass of air moves northward and heavy, warm rains are produced. The rain added to the melting snow leads to large quantities of water rushing toward lower elevations. When the ground is frozen, little water is absorbed by the soil. Since tributaries and main channels are not large enough to handle an excessive quantity of water, the water overflows their banks.

A memorable flood occurred in January and February of 1937, when the Ohio River flooded and caused great damage all along its path, especially in Louisville, Kentucky. The tally sheet showed 436 people dead and more than half a billion dollars in damage. In 1965 heavy rain for several days over eastern Colorado produced tremendous flooding over the banks of the Platte and Arkansas Rivers. Damage was well over $200 million. During the three spring months of 1973 most of the states forming the drainage basin of the Mississippi River had rainfall amounting to 40 cm to 60 cm, one and a half to two times the normal. As a result, there was widespread and extremely costly flooding of cities, towns, and farmland along the Mississippi.

As was noted in Chapter 9, hurricane winds produce surges of ocean water that wash over low-lying coastal areas causing enormous destruction and, in some cases, tremendous loss of life. Usually, the rain of a hurricane is not sufficiently intense and long-lived to cause major floods. Occasionally, however, especially when a hurricane moves over hilly and mountainous terrain, the quantity of rain can be overwhelming. In August, 1969, hurricane Camille, in its dying stages, moved over the rugged Appalachian terrain of Kentucky, West Virginia, and Virginia. Huge quantities of rain fell—one station in Virginia reported more than 78 cm (31 in.) in less than 5 hours. In June, 1972, hurricane Agnes moved northeastward along the coast of the United States and caused massive flooding in Virginia, Maryland, Delaware, Pennsylvania, and New York. In Pennsylvania 24-hour rainfalls exceeded 30 cm (12 in.) at a number of stations around Harrisburg, the state capitol, with one location measuring up to 48 cm (19 in.). Damage attributed to this hurricane, mainly as a result of flooding, amounted to *$3.097 billion*; the toll in human lives was 117 in the United States and 7 in Cuba.

Not much more need be said to emphasize the point that too much rain in too short a time produces floods that cost lives and property. In varying degrees they occur every year.

CLIMATIC CLASSIFICATIONS*

One of the most widely used classifications of world climates was devised in Austria by Wladimir Köppen. It was first published in 1918 and has been modified several times since then. It is based mostly on annual and monthly averages of temperature and precipitation. Köppen recognized that the effect of precipitation on plant development also depended on *evapotranspiration*, the amount of moisture lost through evaporation and transpiration. Clearly, in a hot region a given amount of precipitation is less effective than in a cool region. He devised formulas for combining precipitation and temperature to produce an index of precipitation effectivenenss. Köppen was confronted with the fact that the spacing of observing stations was inadequate to delineate climatic regions, a situation that still exists. To overcome this problem, he used the distribution of natural vegetation to indicate the boundaries of various climatic regimes.

In Köppen's classification there are 5 principal climatic groups that are supposed to correspond with 5 classes of natural vegetation: (A) tropical rainy climates, (B) dry climates, (C) rainy climates with mild winters, (D)

*Readers interested in a detailed discussion of this subject may wish to consult J. R. Mather, *Climatology: Fundamentals and Applications*, McGraw-Hill, 1974.

rainy climates with cold winters, and (E) polar climates. Groups A, C, and D are further subdivided into one of 3 subsets depending on whether there is no dry season (f), dry summer season (s), or dry winter season (w). Group B climates are divided into 2 subsets depending on whether they are semiarid (steppe-type) (S) or arid (desert-type) (W). Group E climates are subdivided according to whether the region is tundra (T) or an ice cap (F). Since the combinations As and Ds are rare, there remain a total of 11 climatic types for the world, each represented by a series of letters.

Köppen's classification has been criticized by geographers for a variety of reasons. It has been argued that the use of precipitation and temperature alone does not adequately account for precipitation effectiveness. Another weak feature is that it uses the same formulas at high and low elevations. Finally, it has been noted that Köppen's procedures do not satisfactorily account for transitions between one climatic regime and a neighboring one. It implies an abrupt change from one climatic type to another.

In 1931 C. Warren Thornthwaite, an American geographer, proposed a new climatic classification. He subscribed to Köppen's ideas that vegetation responds to climate and therefore that plants can be used as meteorological indicators. A unique feature of Thornthwaite's scheme was that he measured precipitation effectiveness by dividing monthly precipitation (P) by total monthly evaporation (E), a quantity called the P/E ratio. The sum of the 12 monthly ratios was named the P/E index. On this basis, Thornthwaite defined 5 *humidity provinces* associated with characteristic vegetation (Table 11-3). Each of the humidity provinces was subdivided into 4 subsets depending on seasonal distribution of precipitation. In addition, mean monthly temperatures were used to establish 6 temperature provinces.

Table 11-3   Humidity Provinces Defined by Thornthwaite

Humidity province	P/E index	Characteristic vegetation
Wet	128 or more	Rain forest
Humid	64–127	Forest
Subhumid	32–63	Grassland
Semiarid	16–31	Steppe
Arid	Less than 16	Desert

Although there are some 120 combinations of the 3 variables specified by Thornthwaite, from his examination of the available data he specified 32 climatic types around the world. This is about 3 times as many types as had been proposed by Köppen. One of the serious shortcomings of Thornthwaite's procedure is the paucity of data on monthly evaporation that are needed for the calculation of the P/E ratios.

In 1948 Thornthwaite proposed a new method for classifying climates and modified it in subsequent years. The method depends only on climatological data and it puts primary emphasis on a moisture index that compares the demand for water by the natural vegetation with the available supply furnished by precipitation. The demand is expressed in terms of the *potential evapotranspiration* (PE), a quantity that refers to the maximum amount of moisture that would be transferred *if the water were available*. Thornthwaite developed a simple formula for calculating PE from a knowledge of air temperature and the length of the day. The greater the amount of precipitation in relation to the potential evapotranspiration, the higher the moisture index and the more humid the climate. In arid regions the value of the moisture indices are negative because precipitation is less than the PE.

In the later Thornthwaite classification, he also used factors taking into account the thermal character of the region and seasonal variations of temperature and of moisture availability.

Various climatologists, among them Hermann Flohn in Germany, have suggested that purely descriptive classifications such as those of Köppen and Thornthwaite are inadequate. Flohn argued that a satisfactory classification should take into account the causes for the climate. In 1950 Flohn proposed a scheme that takes the general circulation of the atmosphere as one of the starting points in explaining the climatic patterns over the earth.

In Chapter 6 a discussion was given of the general circulation of the atmosphere. The patterns of pressure and air motions over the earth account for many aspects of the distribution of climates. As noted earlier, in a high-pressure region subsidence generally prevails. It produces a temperature inversion as the sinking air is warmed adiabatically. The warming leads to a reduction in the relative humidity. These factors inhibit the growth of clouds and precipitation. As a result, under the semipermanent anticyclones around the globe there are deserts, especially under the eastern parts where subsidence is most pronounced.

The rainfall map in Fig. 11-11 shows that extensive deserts exist, at lower latitudes, over the western parts of the continents and other regions where anticyclonic motions predominate. Examples are the Sonora desert in the southwestern United States and Mexico, the extensive desert along the west coast of South America and North Africa. It is important to recognize that the areas of deficient rainfall extend over the adjacent oceans. This should be recalled if you hear it proposed that the construction of an artificial lake might increase rainfall in a desert region. A scarcity of rainfall generally can be attributed to the predominance of subsidence rather than the absence of a nearby source of water vapor.

Wet climates are found in regions that have strong, persistent ascending air motions, especially if the air is warm and humid. These conditions are commonly found along the intertropical convergence zone. Mountain ranges force air upward as well as acting as sources for convective clouds when their slopes are warmed by the sun. When the air moving up the mountain slopes is warm, humid, and unstable, spectacular amounts of rain can fall. Table 11-1 gives some examples.

Figure 11-11 illustrates how the Rocky Mountains along the west coast of the United States and Canada experience heavy precipitation on their windward sides. The prevailing winds are westerly and they carry moist air from over the Pacific Ocean. Note, however, that desert conditions are found on the leeward side of the front range of mountains. For example, in eastern Washington precipitation is less than 25 cm per year.

The rainfall minimum downwind of a mountain is known as a *rain shadow*. As the air in the westerlies moves up the western slopes of the mountains, its temperature falls and its relative humidity increases. Clouds develop; rain and snow fall to the ground. When the air then passes over the mountains and sinks, it warms adiabatically. Since water substance has been removed, the descending air is left much drier than when it passed over the coastline. Since clouds and precipitation do not occur readily in the subsiding air, the result is a desert.

The seasonal climates of an area are indicated by the seasonal character of the general circulation. As shown in Fig. 6-3, the average air currents in the middle of the atmosphere meander around the earth in a series of shallow troughs and rideges. On the average, regions with winds from a northerly direction are colder than regions with winds from a southerly direction. There also tends to be more precipitation over regions under pressure troughs. As a result, the eastern parts of the United States are cold and wet in an average winter.

In certain years the patterns of troughs and ridges in the pressure field differ drastically from the average ones shown in Fig. 6-3. For example, in the winter a deep 500-mb trough may become established over the western part of the United States. In such a circumstance, the western states would be abnormally wet and cold. At the same time the eastern states would be abnormally warm and dry.

Sometimes the entire pattern of pressure troughs and ridges over the Northern Hemisphere shifts longitudinally. When this happens, weather abnormalities occur in a great many places. For example, if it is unusually warm and dry in winter in New England, it should not be surprising to learn that western Europe is abnormally cold or that parts of the Soviet Union are abnormally warm and dry.

The reasons for shifts in atmospheric circulation leading to abnormal seasonal or annual weather still are not understood. Some scientists suggest that they are more or less random events. On the other hand, it has been proposed by various meteorologists, among them the well-known American meteorologist, Jerome Namias, that abnormalities of ocean temperature have a controlling influence on large-scale atmospheric motions over periods of months.

CLIMATES
OF THE EARTH

It is known that ever since the earth developed into a planet with land, water, and air there have been slow but continuous changes in climate. They appear in geological records as glacial epochs interspersed with long-lasting warm intervals. Widespread glaciation occurred during the Precambrian, Devonian, and late Carboniferous periods some 700, 400, and 330 million years ago, respectively, and more recently, during the Pleistocene epoch which started 700,000 to 1 million years ago.

During the glacial epochs large regions of the earth were covered periodically with sheets of ice that advanced equatorward, in extreme cases reaching to latitudes of 40° over the continents.

As might be expected, the recent past has been of particular interest to many scientists. The Pleistocene epoch has been studied extensively by paleoclimatologists who have used a variety of techniques for dating fossils and relics and for estimating temperatures during their formation.* Unfortunately, a completely reliable method for dating events between 150,000 and 10,000,000 years old still does not exist. This fact, coupled with contradictory findings in various parts of the world, has prevented the construction of a consistent geological calendar over the Pleistocene or even a major fraction of it. Nevertheless, the available evidence does allow some estimates of the major climatic occurrences.

The Pleistocene epoch experienced at least four major ice ages during which the average temperature of the earth was about 6°C below today's average value. Each glacial age lasted about 100,000 years. The last one called the *Wisconsin Age* started about 120,000 years ago and ended only 10,000 years ago. The interglacial ages during the Pleistocene epoch were short warm periods with temperatures of the earth averaging about 3°C higher than those of today.

Looking more closely at the recent records, researchers have found that during the Wisconsin Age there were four glacial advances. One of the best documented, which started some 23,000 years ago, was accompanied by temperatures over the Central Atlantic Ocean, the Caribbean, and Europe which were about 10°C colder than present-day temperatures. The climate of the United States was wet and cool with glaciation over the central part of the country extending as far south as Iowa and Nebraska.

Data on the climate over the last 10,000 years has been fairly extensive. Table 11-4 contains a small sampling of certain important features of the climate. Of particular note is the interval 5600 B.C. to 2500 B.C. when the atmosphere was 2°C to 3°C warmer than it is now and conditions were moist,

---

*For a discussion of geological dating techniques, see D. L. Eicher, *Geologic Time*, Prentice-Hall, 1968.

Table 11-4   A Brief Chronology of the Climate of the Last 10,000 Years

Dates	Region	Climate
9000–6000 B.C.	Southern Arizona	Warm and arid
7800–6800 B.C.	Europe	Cool and moist, becoming cool and dry by 7000 B.C.
6800–5600 B.C.	North America, Europe	Cool and dry, with possible extinction of mammals, particularly in Arizona and New Mexico
5600–2500 B.C.	Both hemispheres	Warm and moist, becoming warm and dry by 3000 B.C. (Climatic Optimum)
2500–500 B.C.	Northern Hemisphere	Generally warm and dry with periods of heavy rain and intense droughts
500 B.C.–0 A.D.	Europe	Cool and moist; glacial maximum in Scandinavia and Ireland between 500 B.C. and 200 B.C.
330	United States	Drought in the Southwest
600	Alaska	Glacial advance
590–645	Near East, England	Severe drought in the Near East, followed by cold winters; drought in England
673	Near East	Black Sea frozen
800	Mexico	Start of moist period
800–801	Near East	Black Sea frozen
829	Africa	Ice on the Nile
900–1200	Iceland	Glacial recession (Viking period)
1000–1011	Africa	Ice on the Nile
1000–1100	Utah	Snowline 300 m higher than today
1200	Alaska	Glacial advance
1180–1215	United States	Wet in the West
1220–1290	United States	Drought in the West
1276–1299	United States	"Great Drought" in the Southwest
1300–1330	United States	Wet in the West
1500–1900	Europe, United States	Generally cool and dry; periodic advances in Europe (1541 to 1680, 1741 to 1770, and 1801 to 1890) and North America (1700 to 1750); drought in the southwestern United States from 1573 to 1593
1880–1940	Both hemispheres	Increase of winter temperature by 1.5°C; drop of 5.2 m in the level of the Great Salt Lake; Alpine glaciation reduced by 25% and Arctic ice by 40%; rapid glacial recession in the Patagonian Andes (1910–1920) and the Canadian Rockies (1931–1938)
1942–1960	Both hemispheres	Temperature decrease, mostly in Northern Hemisphere, and halt of glacial recession
1961–1975	Northern hemisphere	Little change in the average air temperature

Source: Mostly W. D. Sellers, *Physical Climatology*, University of Chicago Press, 1965.

especially over North Africa and the Middle East. This period is known as the *Climatic Optimum* because conditions were favorable for development of plants and animals.

The most recent cold period, known as the *Little Ice Age*, occurred fairly recently, from 1500 A.D. to 1900 A.D. During those 400 years it was generally cool and dry and there were equatorward advances of glaciers and sea ice. This cool period is clearly shown in Fig. 11-15 which depicts annual temperatures in Iceland over the last millenium.

Near the beginning of this century a pronounced warming began over most of the Earth with most noticeable temperature changes in the northern part of the Northern Hemisphere. Sea-level temperatures averaged over the Northern Hemisphere are shown in Fig. 11-16. They not only show very large year-to-year fluctuations, but also clearly exhibit an upward trend from about 1880 to 1940. A five-year running mean would show a temperature increase of about 0.6°C. The warming was accompanied by a poleward displacement of the edges of sea ice, a retreat of glaciers, and a small increase in sea level. As shown in Fig. 8-28, in extreme cases of glacier growth or shrinkage, sea level could change by tens of meters. Imagine the devastation in the coastal cities of the world if sea level were to rise even a few meters.

In about 1940, the lower atmosphere began a cooling trend, especially in the Northern Hemisphere, and it continued until the early 1960's. As air

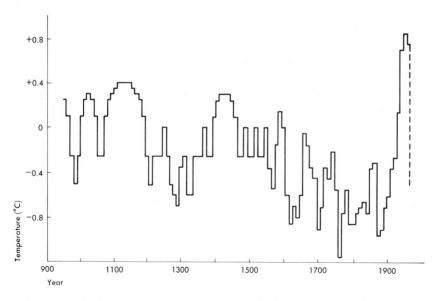

Figure 11-15 Ten-year average temperatures in Iceland over the past millenium. The dashed line on the right indicates the temperature decline in the 1961–1971 period based on additional data. *Courtesy* P. Bergthórsson, Icelandic Meteorological Office, Reykjavík and R. Bryson, University of Wisconsin, Madison.

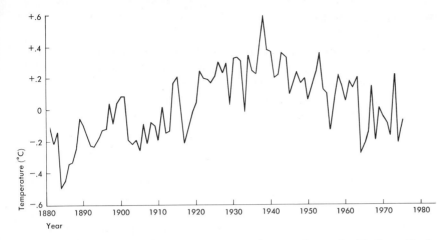

Figure 11-16 Mean annual air temperature near the ground over the Northern Hemisphere. The data from 1880 to 1960 were provided by M. I. Budyko; the remainder were provided by J. K. Angell and J. Korshover. *Courtesy* J. Murray Mitchell, Jr., National Oceanic and Atmospheric Administration.

temperatures decreased, sea ice thickened and advanced equatorward, glaciers advanced, and the threat of rising ocean waters was dispelled. Since the 1960's, as shown in Fig. 11-16, average temperatures over the Northern Hemisphere have generally been level or on a slow warming trend.

The reasons for these major climatic fluctuations over periods of tens of years still are not known. Atmospheric scientists are particularly interested in learning if these fluctuations are *anthropogenic*, that is, if they can be caused by human activity, such as the emissions of pollutants into the environment. A central problem facing climatologists is the development of the necessary understanding to make reliable predictions of future climatic conditions.

HYPOTHESES
ON CLIMATE CHANGES

It is widely acknowledged that a single hypothesis cannot account for the observed climatic changes over the entire history of the earth. Various factors are recognized as being responsible for climate changes depending on the time scales involved. It is generally agreed that climate changes over periods of 100,000 to a billion years can be explained in terms of continental drift.* The various continents very slowly, but for very long periods, move with respect to one another. A comparison of the east coast of South America

*See S. P. Clark, Jr., *Structure of the Earth*, Prentice-Hall, 1971.

and the west coast of Africa indicates how they could be fitted together like parts of a jigsaw puzzle. Geologists have developed a body of knowledge dealing with the dynamics of the earth and has given it the name *geodynamics*.

There is increasing evidence that climate changes over periods of 10,000 to 100,000 years are caused by orbital variations of the earth around the sun. This theory for the formation of ice ages was proposed in about 1930 by the Serbian scientist Milutin Milankovich. He noted that there are slow, continuous changes of the time when the earth is closest to the sun in such a fashion that the Northern Hemisphere should experience warm summers and cold winters at periods of about 20,000 years. The angle that the plane through the earth's equator makes with the plane of the earth's orbit around the sun varies with a period of about 40,000 years. Finally, the eccentricity of the earth's orbit (that is, the deviation from a circular orbit) varies with a period of about 95,000 years. Recent observational data show good correlations between inferred temperatures in the Northern Hemisphere and incoming solar radiation changes attributed to the factors of the Milankovich theory.

For hundreds of years astronomers and other sun watchers have kept records of dark areas on the sun called *sunspots*. Many studies have been made relating characteristics of the general circulation and climate to sunspot number. One of the best-known analyses was first reported in 1951 by Hurd Willett, then at the Massachusetts Institute of Technology. He suggested that the modulations in climate were not caused by gradual variations in solar radiation, but rather by the more irregular ones associated with solar eruptions. He compared fluctuations in climate over a 200-year period with the frequency of sunspots observed over the same interval. On the basis of this admittedly short record, Willett suggested that there is an 80-year cycle in the climate associated with a similar cycle in sunsport occurrence.

It is well known that sunspot numbers vary with a period of about 11 and 22 years. Many meteorological factors such as rainfall sometime exhibit variations over the same periods. In 1977, the well-known American climatologist, J. Murray Mitchell, and his associates reported new evidence relating sunspot frequency to drought occurrences over the western United States.

During certain lengthy periods sunspots virtually disappear from the face of the sun. For example, there were almost none from about 1645 to 1715 A.D. This 70 year period occurred near the middle of the Little Ice Age.

Unfortunately there still is no acceptable explanation for how a change in sunspot number can influence the circulation of the atmosphere. Until such an explanation is found, it is premature to claim that changes in sunspot numbers cause changes in the climate.

There is a variety of "nonastronomical" hypotheses for global climate changes. It has been proposed that volcanism might account for the formation of ice ages. Volcanic eruptions introduce massive quantities of particulate matter into the atmosphere (Fig. 2-5). Such particles remain in the strato-sphere for a year or two and reduce the quantity of solar radiation reaching the ground. Recent work shows that particles in the atmosphere can cause *net* warming or cooling depending on the characteristics of the particles, the altitudes at which they exist, and the reflectivity of the underlying surface.

It has been proposed that changes in past climates may have been brought about by variations in the carbon dioxide concentration in the atmosphere caused by volcanic eruptions. Most scientists believe that unre-alistically large changes in the concentration of carbon dioxide would have been needed to explain temperature changes known to have occurred over geological time.

Unfortunately, a critical evaluation of the many theories of climatic change is hampered by an inadequate understanding of the physical mecha-nisms of the general circulation.

CLIMATIC CHANGES
OVER THE LAST CENTURY

Geophysicists have recently focused a great deal of attention on the variations of the earth's temperature over the last century (Fig. 11-16). It is particularly important to learn if the warming and cooling can be attributed to human activity.

There is no doubt that man has changed the environment in many ways. As noted in Chapter 2, the composition of the atmosphere has been altered somewhat. There has been a significant increase in the concentration of carbon dioxide, and to a lesser extent, increases in the concentrations of certain other gases and particulates. Oil spillage from ships and drilling operations have produced a film of oil over much of the ocean surface. Forests and fields have been cleared and replaced by highways and cities. Large quantities of heat are put into the atmosphere, rivers, and lakes. It does not appear that environmental alterations other than the additions of gases and particles to the atmosphere are influencing the global climate.

As noted in Chapter 2, the rise of carbon dioxide concentration in the atmosphere over the last century can be ascribed mostly to the burning of fossil fuels. Roughly half of the released carbon dioxide is retained in the atmosphere and the remainder is absorbed by the oceans or is taken up by plants. It has been estimated that by the end of the twentieth century the carbon dioxide concentration will be about 380 ppm. Projections into the future, assuming that the major energy sources will be coal and nuclear energy, indicate a concentration about twice the 1900 value by the middle of the next century.

If no other factors were operating, the continual increase of carbon dioxide in the atmosphere would be expected to cause a warming of the atmosphere because the carbon dioxide would absorb some of the terrestrial infrared radiation that would normally escape to outer space through the "windows" in the water vapor absorption spectrum (see Fig. 3-4). But carbon dioxide is not the only factor affecting atmospheric temperatures.

From about 1940 to about 1960 temperatures over much of the earth, especially in the northern part of the Northern Hemisphere, had a downward trend and have changed little since then (Fig. 11-16) even though carbon dioxide increased through this period. It has been speculated that this cooling might be attributed to an increase of particulates in the atmosphere. A careful analysis of the problem shows that if all other factors were constant and particulate concentrations in the atmosphere were increased, air temperatures near the ground would not necessarily decrease.

The available observations of atmospheric particles paint an uncertain picture of the trends over the planet as a whole. The curves shown in Fig. 2-5 indicate that over Mauna Loa, Hawaii, the values of atmospheric turbidity increase following large volcanic eruptions, but the particles are cleared out of the atmosphere over a period of several years. On the other hand, measurements over the North Atlantic Ocean indicate that the particle concentration nearly doubled between 1907 and 1970. The same observational technique showed no change in particulate loading over the South Pacific.

In summary, the limited data that exist do not readily support the notion that atmospheric cooling since the 1940's can be attributed simply to increased particulate pollution.

It has been suggested that as the population increases, the heat released into the atmosphere by motor vehicles, industrial plants, the heating and cooling of buildings, and other societal activities might have a significant impact on the global climate. Analyses to date indicate that man-made heat releases through the end of this century and well into the next are not likely to be significant and certainly not as important as the increasing concentrations of carbon dioxide.

On a local scale, heat released from large urban areas certainly can have noticeable effects on the local climate. As noted earlier, temperatures are higher in cities, mostly because of the heat retained in the streets and buildings, in part because of the heat generated and released within the city.

Theoretical studies of the general circulation make it abundantly clear that climate variations cannot be explained by examining one or two of the variable properties of the atmosphere and the land and water below it. As was mentioned earlier, the general circulation is a complex mechanism that has many feedbacks. A change in one feature causes changes in others that react with the first, and so forth. For example, an increase in temperature may cause an increase in evaporation, a rise in relative humidity, and atmospheric instability. The result could be the formation of clouds that are better reflectors than the earth's surface, a decrease in the amount of solar radiation reaching the lower atmosphere, and a cooling of the lower atmosphere.

An advanced model of the general circulation of the atmosphere has been used by Syukuro Manabe and Richard T. Wetherald of the Geophysics Fluid Dynamics Laboratory at Princeton University to evaluate the effects of a doubling of the carbon dioxide concentration in the atmosphere. The calculations indicate that the consequences would be about a 2°C increase in air temperature near the ground. In view of the fact that the earth's average temperature difference between periods of ice ages and the interglacial periods amounted to less than 10°C, a change of 2°C would be of great significance. The less than a 1°C change over this century shown in Fig. 11-16 has had major consequences on the atmosphere and oceans.

Although the Manabe-Wetherald model was the best available in the 1970's, it still did not adequately account for interactions between the atmosphere and the oceans. For this reason, their results cannot be regarded as proven facts. Nevertheless, in view of the steady increases of carbon dioxide which are expected to continue into the distant future, the Manabe-Wetherald calculations serve to alert us of the crucial need to establish the facts as soon as possible. This requires the development of much better theoretical models of the global climate than exist today.

CHAPTER 12

Applications

THE WEATHER AND CLIMATE affect people and their possessions in a great many ways; some are obvious, but others are so subtle that they may not be recognized. For example, there is considerable evidence that the weather can have important physiological influences on some individuals. Many have reported correlations between asthmatic, arthritic, or sinus troubles and some aspects of the weather. These claims have some validity even though they cannot be explained adequately.

When a strong foehn wind is blowing (Chapter 4), certain people experience distinct psychological as well as physical reactions. The latter are attributable to the high temperatures and extreme dryness. It is not clear why such winds are accompanied by increases in instability, headaches, and suicides.

There have been studies of how decisions of great men were influenced by the state of the atmosphere. Of course, when you set out to find relationships, you often succeed even if real ones do not exist. Nevertheless, it is certainly true that sometimes the weather makes you feel good while on other occasions it has the opposite effect. A warm, sunny day after a long, cold winter makes the world seem brighter.

History abounds with examples of the important role played by the weather in battles that affected the fate of nations. The bitter, winter days of 1776 put great strains on George Washington and his poorly equipped army. They survived and went on to victory. The French under Napoleon and the German army in World War II found the Russian winter an unbeatable enemy.

When airplanes became major factors in warfare, the weather became even more important than ever. As a result, the U.S. Air Force has its own weather organization to meet its special needs. This sort of arrangement is found in many other countries as well.

Weather is vital in peace as well as in war. In early chapters the disastrous effects of violent weather were mentioned. Too much rain causes flooding; too little causes droughts and serious losses to agriculture and many other water-related activities. Snow over mountain watersheds represents stored water to serve agricultural, municipal, and industrial needs. Too much snow over urban areas can bring a city to a halt (Fig. 12-1); deep snow over grazing lands can cut off feed and water to isolated animals.

Figure 12-1 Washington, D.C. immobilized by a snow storm. *Courtesy* National Oceanic and Atmospheric Administration.

Strong winds can cause physical damage to plant life as well as buildings. Persistent, strong winds increase transpiration and evaporation. In rain-deficient areas the winds put even greater strains on available water. In extreme cases, they can lead to significant reductions in crop yields.

Long periods of low humidity during the period of active plant growth contributes to desiccation of soil and plants. On the other hand, low humidities are needed during the drying of corn, wheat, and other grains just before harvest. If harvesting is delayed because of inadequately dried grains or wet fields that hinder the passage of machines, the usual result is a reduction in crop yield.

One could cite many other activities in which the weather plays a crucial role. They range from the baseball games postponed because of rain or wet grounds to the town whose electric power is lost because of a lightning strike to a critical transformer. Some businesses thrive on the threat of inclement weather, for example, manufacturers of lightning rods, umbrellas, and raincoats; on the other hand, manufacturers of bathing suits and golf clubs long for sun and warmth.

We can be sure that over the ages man has sought to learn the nature of the atmosphere and use that information to improve his circumstances. As better techniques for observations and communications have developed and as science and technology have advanced, we have been using the accumulating knowledge to ever-increasing degrees.

USE OF WEATHER
AND CLIMATIC DATA

Available climatological data have a great many applications. For example, when building airports it is essential that the runways be aligned along the direction of the prevailing winds because it is preferable that airplanes take off and land into the wind. A diagram such as the one shown in Fig. 12-2, assembled from past observations and called a *wind rose*, shows that the surface wind is most often from the southeast in this particular area and it tells an airport designer the direction he should use when laying out the runways.

Successful farming depends on the intelligent use of climatological information. The planting of crops must take into account the dates when temperatures are likely to fall below the freezing point as well as the length and characteristics of the growing season. Frost-sensitive vegetables cannot be put into the fields until the likelihood of a killing frost is very small. At the same time, plant scientists can specify the number of days of above-freezing temperatures needed for most crops and therefore put limits on the kinds of vegetation that can be expected to thrive in any climatological regime.

Everyone knows that weather extremes can damage crops and reduce yields, but it is not so well understood that the rate of development of plants during various stages of growth depends on meteorological conditions. The study of how climate affects periodic events in the lives of plants and animals is called *phenology*. The dates of planting, germination, emergence from the soil, budding, flowering, ripening, and harvesting are important phenologically. The responses of the plants are influenced by the westher, especially temperature, rainfall, and humidity, before and at the time of occurrence of the phenological events.

The interested reader should consult the literature on agricultural climatology for discussions of this subject. We shall illustrate the importance of weather and climate by noting that the productivity of any crop depends not only on the length of the growing season but also on the quantity of energy available to the vegetation during the season. This can be measured

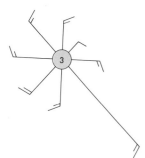

Figure 12-2   A surface wind rose for Tucson, Arizona, showing the relative frequency of various wind directions. The length of the arrow is proportional to the percent of occasions with wind from that direction. The number in the center gives the percent of observations with calm winds. The average wind speed is shown by the barbs. Each full-length barb equals 2 m/sec. Note that 1 m/sec = 2.24 mi/hr.

by summing the number of *growing degree-days* during the period between planting and harvesting. A degree-day for any crop is defined as the departure of the average daily temperature above the *zero temperature* of that crop. The zero temperature is the minimum temperature for the growth of any plant. According to Howard J. Critchfield*, some typical zero temperatures are 40°F, 43°F, 45°F, and 50°F for peas, oats, potatoes, and sweet corn, respectively. For example, if the average daily temperature were 50°F, the number of degree-days on that day would amount to 10, 7, 5, and 0, respectively, for the four listed crops.

Starting from the day of planting, the number of degree-days is accumulated day by day. For each crop there is a minimum total of degree-days necessary to allow it to pass successfully from planting to harvesting. If the climatological data show that, in any region, the available number of degree-days is fewer than the necessary amount, the crop under consideration should not be planted. In low-latitude areas having long, hot summers the annually available number of degree-days may be large enough to allow two or, in extreme cases, three crops on the same land during a calendar year.

In summary, it can be said that by knowing the climate of a region and the phenology of crops likely to be suitable there, it is possible to make sound judgments on which crops to plant and to establish schedules for planting, fertilization, spraying, and harvesting.

Incidentally, during the winter accumulated degree-days are used by power companies and air-conditioning engineers to measure the quantities of fuel needed for heating purposes. In this application the number of *heating degree-days* on any day is the number of degrees by which the average daily temperature falls below a standard temperature, which in the United States is 65°F. For example, if the average daily temperature is 40°F, that day would have had 25 degree-days. The number can be accumulated for any period such as a month or a year. As would be expected, the annual total of degree-days at any place is a reflection of the annual average temperature. In southern Arizona and Florida the average annual heating degree-days amount to less than 1000; in San Francisco the number is about 3500; in Chicago it is about 6500; and it increases greatly at more northerly latitudes through Canada and Alaska. Clearly, the greater the number, the greater the fuel requirements and the need for large fuel storage facilities and distribution systems.

As noted earlier, strong, persistent winds act to reduce agricultural yields. Where these persistent winds exist, rows of trees, bushes, or artificial barriers can be installed to produce a *shelterbelt* in order to reduce wind speeds.

A knowledge of the climate is also important to ranchers, dairymen, and swine and chicken growers. Most animals respond in a distinctive

*Howard J. Critchfield, *General Climatology*, 3rd ed., Prentice-Hall, Inc., 1974.

way to temperature and humidity conditions. When it is too hot and humid, the production of domestic animals decreases. They eat less and produce fewer eggs or less milk; cattle and hogs increase in weight slowly. A modern farmer uses a knowledge of the climate to plan shelters for livestock during the period of the year when it is expected to be excessively hot or cold.

The airline industry must be well informed on the climate and weather. This was recognized in the early days of aviation and it accounts for the fact that airline companies played important roles in the establishment of weather stations all over the globe.

The length of runways required for the takeoff of an airplane depends on the properties of the air near the ground. As the air gets warmer and more humid, its density decreases. The density also decreases with increasing altitude. The lower the density, the greater must be the airplane speed in order for the aerodynamic lift forces to be sufficient for the airplane to take off and climb. For this reason, in places such as Phoenix, Arizona, where it is not unusual to experience temperatures above 40°C (104°F) on summer afternoons, long takeoff runs are required when a large, heavily loaded airplane leaves the field. The same is true at Denver which is at an elevation of about 1600 m above sea level. At such high places the takeoffs are even longer when high temperatures and humidities occur.

When planning long-distance flights, aviation meteorologists must take several important factors into account. First of all, it is important to have airplanes flying with the wind whenever possible. The velocity of an airplane over the ground is the sum of the velocity of the airplane through the air plus the velocity of the air with respect to the ground. If an airplane having an air speed of 246 m/s (550 mi/hr) is flying eastward and there is a westerly tail wind of 45 m/s (100 mi/hr), the ground speed of the airplane is 291 m/s (650 mi/hr). On the other hand, if it is flying westward into the wind, the 45-m/sec (100-mi/hr) head wind reduces the airplane's ground speed to 201 m/s (450 mi/hr). In middle latitudes, because the winds are mostly westerly, flying eastward usually takes less time than flying the same distance in the opposite direction. For example, a New York to Paris flight on a Boeing 747 averages about 6 hr, 30 min, while the return flight on the same airplane is scheduled at 8 hr, 15 min.

In efficient flight planning, attempts are made to maximize tail winds and minimize head winds. This practice is important because modern jet airplanes fly at altitudes of 6 km to 12 km (20,000 ft to 40,000 ft), where the strong air currents known as jet streams are found (Chapter 6).

Aeronautical meteorologists must also take into account whether or not there is likely to be air turbulence. When there is strong wind shear, that is, when the wind changes appreciably over a small distance, turbulent air motions result. One can imagine them to be eddies of various sizes in which the air moves in a chaotic fashion. In a sense, they resemble the eddies one sees when a rapidly moving river encounters obstructions and breaks into a turbulent, frothing mass of water. When atmospheric eddies have diameters about the same as the wingspan of the airplane, they cause abrupt movements of the airplane. The plane can experience jarring upward, and downward, and side-to-side jolts. It is an unpleasant experience for pilots and passengers, especially when turbulence is severe. If an inexperienced pilot flying an inadequately stressed airplane encounters violent turbulence, the result can be structural failure and disaster.

It is well known that thunderstorms are regions of turbulence and other hazards such as hail and lightning. They usually can be avoided either visually or by means of a weather radar of the type found in the cockpits of commercial airliners.

The avoidance of clear air turbulence, known as CAT, is much more difficult because usually there are no clouds to serve as markers of the presence of turbulence. When strong winds blow nearly perpendicular to mountain barriers, wave motions occur on the lee side (see Fig. 8-7). Strong updrafts and downdrafts occur with clear air turbulence. A second principal source of CAT is the wind shear in the vicinity of a strong jet stream. Fortunately, meteorologists have learned to recognize many of the wind patterns that are likely to lead to clear air turbulence. In addition, pilots who encounter it radio the news to weather offices and this information is used in laying out the paths of subsequent flights. When moderate or severe turbulence is expected in an area, it is circumnavigated even at the cost of a longer, more time-consuming flight.

In an airplane it is always advisable to wear a seat belt. Occasionally, clear air turbulence is encountered unexpectedly, and in the rare event of severe conditions the belt will keep you in the seat. In general, the turbulence is restricted to a fairly shallow region, and by changing altitude a pilot can find calmer air.

The reader can certainly think of other ways in which a knowledge of the state of the weather and climate can be used beneficially. One place where it has not been used adequately is in city planning. For example, would it not be reasonable to have industrial sites and other producers of environmental pollutants some distance downwind of residential areas? Unfortunately, in many places the zoning boards do not take the climatological characteristics of the region into account.

WEATHER FORECASTING

Weather observations and climatological information are useful in the planning of a wide variety of human endeavors. In many circumstances, accurate predictions of the weather can be of even greater value. It is easy to see that forecasts a month, a week, or a day in advance would be beneficial to businessmen and farmers as well as to the average person concerned with going to or from work or planning a picnic or other outdoor activities.

For certain purposes, a correct forecast even a few minutes in advance is very worthwhile. For example, knowledge that a tornado will strike a building, say a school, in 5 min would permit the occupants to take protective action. At busy airports, such as O'Hare, in Chicago, a precise prediction that a dense fog will close the airport in 5 min would be valuable to the airplane controllers as well as to the pilots. At O'Hare, airplanes sometimes land and take off as frequently as one per minute. If incoming airplanes have to be flagged off and sent to other airports, even minutes can mean substantial reductions in operational costs as well as a contribution to safer operations.

Most weather forecasts are made from several hours to perhaps 2 days in advance. Predictions of daily rainfall and temperature are made regularly up to periods of 3 days and 5 days. For longer stretches of time, up to perhaps a month, general outlooks of average temperatures can be made with some skill, but the longer the period, the less accurate the forecast.

When measuring the skill of a forecaster, it is not enough to ask the percentage of accurate forecasts. This point can be illustrated by examining how well one can forecast rain in Los Angeles during the dry months of July and August. Over the period 1960 to 1976 measurable rain (greater than 0.25 mm) fell on only 12 days out of a total of 1054 days. Therefore, the chance of rain was 12 out of 1054, or about 1 percent. It is likely to remain close to that in the future. Anyone knowing these facts can predict "no rain" for Los Angeles for every day of July and August, 1990, with confidence of being right close to 99 percent of the time. The percentage is high, but it is not a measure of forecasting skill; it is an indication that there is predictive value in climatological data. As already noted, such information is very worthwhile, but to demonstrate skill a weather forecaster should be able to do better than merely use climatology. In Los Angeles, for example, a good forecast would accurately predict those rare days in July and August on which rain does occur.

Another weather forecasting technique that in many places yields results that are better than those given by pure chance is called *persistence forecasting*. The procedure is to predict for tomorrow what happens today. For example, if it rains today, predict that it will rain tomorrow. This scheme is better than flipping a coin because in many places rainy and dry days run in series; there may be 4 dry days followed by 3 days with rain and then another stretch of dry days. Persistence forecasting has the obvious disadvantage that it always misses on days when the weather changes.

For very short periods—those of a few hours or less—an assumption of persistence is a reasonable one. For example, if it is raining at a station, particularly from a cyclonic storm, it would be expected that it would still be raining an hour later. As the period increases, however, the accuracy of a persistence forecast would diminish rapidly.

Meteorologists have devised many schemes for evaluating forecasting skill. The better ones compare the accuracy of the forecasts with climatological expectations. Climatological data can be used to predict rain for a series of days in such a fashion that the frequency of forecasted rainy days is about the same as occurred in the past. For example, imagine that in a particular city the climatological records show that, on the *average*, it rains 10 days in June. A bowl could be filled with 10 white balls and 20 black ones. After the bowl is shaken, a blindfolded person could take out the balls randomly one at a time, with each successive ball indicating the weather for each successive day. If the first ball removed were white, it would be predicted that rain would occur on June 1. If the next ball were black, no rain would be predicted for June 2, etc. This procedure leads to a series of forecasts based on climatological expectations. A weather forecaster using whatever information and procedures are available also can predict the occurrence of rain for each day in a particular June. The results of the two sets of forecasts could then be compared with the actual observations of days with rain. One simple procedure for estimating forecast skill would be to take the ratio of the fraction of correct forecasts by means of the two procedures. For example, if the weather forecaster were correct 23 out of 30 days and the climatological prediction were correct on 19 days, the skill could be represented as $23/30 \div 19/30 = 0.77 \div 0.63 = 1.22$. The greater the value of the ratio, the more skillful the forecasts.

The procedure used for making weather forecasts is to examine the most up-to-date observations of the weather and the state of the atmosphere and then predict the changes expected to occur. As would be expected, the longer the forecast period, the greater the amount of information needed at the outset. This point is illustrated in Fig. 12-3. When forecasting conditions at the point p in the middle atmosphere 12 hours to 36 hours in the future, the inertia of the atmosphere is sufficiently large so that it is enough to consider only conditions over a small part of the earth. Figure 12-3 shows that observations at one altitude over a latitude zone of about 40° would give adequate information. For a 3-day forecast, initial observations must be taken over most of the hemisphere and from near the ground to the base of the stratosphere. When the forecast period is 3 days to 5 days long, it is necessary to know initial conditions over a hemisphere and take into account exchanges of energy, water vapor, and momentum between the atmosphere and the earth's surface, especially over the oceans. If a forecast is extended to 2 weeks, initial conditions must be known over the entire earth and to a depth of several meters in the oceans.

Until the 1960's weather forecasters depended largely on subjective analyses of surface and constant pressure charts, particularly the one at

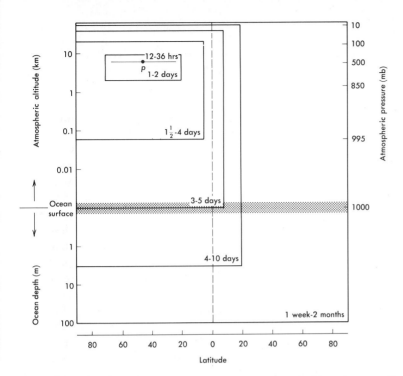

Figure 12-3 Initial data over the indicated regions are required in order to predict conditions at point *p* for the indicated periods. The shaded area is the ocean–atmosphere interface zone. *From* J. Smagorinsky, *Bulletin of American Meteorological Society*, 1967, **48**: 89–93.

500 mb. By means of extrapolations and empirical rules, the location of fronts and the positions and intensities of low- and high-pressure systems were predicted. Once this was done, winds could be obtained by knowing pressure gradients, and the weather (that is, clouds, precipitation, and temperatures) was predicted mostly on the basis of the air mass characteristics, wind patterns, and frontal locations. Some people who used these techniques produced forecasts of amazing accuracy. Unfortunately, this approach was as much an intuitive art as a science. Outstanding forecasters did not seem to be able to teach others how to do it.

Since the late 1940's there has been a steady growth of the use of mathematical models for weather prediction. These procedures have been made possible by advances in the formulation of mathematical models of the atmosphere and the development of high-speed computers. The general procedure is similar to the one used to develop theoretical models of the general circulation (see Chapter 6). A series of equations is used to specify changes of the atmosphere with time. They take into account air motions, temperatures, humidities, evaporation at the ground, clouds, rain, snow, and interactions of the air with land and oceans. For daily weather forecasting,

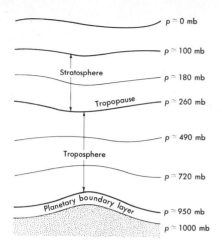

Figure 12-4 The six layers of the atmosphere used by the National Weather Service in a baroclinic, numerical weather-prediction model. *From* F. G. Shuman and J. B. Hovermale, *Journal of Applied Meteorology*, 1968, **7**: 525–547.

the National Weather Service uses a numerical model in which the atmosphere is divided into six layers (Fig. 12-4). In experimental programs on the use of numerical models to predict the state of the atmosphere 1 to 2 weeks in advance, the atmosphere has been divided into as many as 11 layers.

Since the mathematical model calculates changes of the atmosphere with time, it is necessary to start with a reasonably complete and accurate knowledge of the initial state of the atmosphere. Observations of the atmosphere are made twice a day at 0000 GMT and 1200 GMT by means of radiosonde stations over most of the continental areas of the earth and from some islands and ships at sea.

Radiosonde data are augmented by means of radiometric observations from satellites that allow calculations of temperature at the surface and its distribution with height. Satellites also yield data on atmospheric humidity and cloud cover. These quantities are used in the construction of maps showing the state of the atmosphere.

At the National Meteorological Center near Washington, D.C. the process of weather analysis and prognosis is done almost entirely by computers. Observations of temperature, humidity, pressure, and wind velocity are collected; then automated techniques are used to draw maps showing the patterns of these quantities. The computers use the mathematical model to calculate pressure, temperature, and wind distributions at a grid of regularly spaced points at various levels in the atmosphere.

The process involves a stepwise computing procedure. Starting with known conditions, calculations are made of the changes of all relevant quantities over a period of about 6 min. This establishes a new set of conditions that serves as the basis for calculating changes over the next 6 min. This procedure is repeated for as many steps as are required to produce prognostic maps 1 day, 2 days, or many days into the future. Because the need for data increases as the forecast period gets longer (Fig. 12-3) and because of the fact that in the computing process, errors in observations are magnified with time, most *operational* numerical predictions are made for periods up to 48 hours. Experimental numerical predictions are made for longer periods, but they are not as reliable as the 1-day and 2-day predictions. As observational coverage increases and as mathematical models improve, the operational forecasts will be extended.

Figure 12-5 presents an example of an operational forecast made by the National Weather Service by means of a mathematical model. The cyclone in western Canada and the deep cyclone over the eastern United States were accurately predicted. The latter storm produced more than 30 cm of snow over the lower Great Lakes region.

The mathematical models also predict vertical air motions and humidity and these data are used to predict precipitation. Figure 12-6 shows the results of a reasonably good forecast. In general, the outcomes of such forecasts are not accurate enough especially during the warm seasons of the year when most rain falls from showers and thunderstorms. The current mathematical models do not yet deal in an adequate manner with small- and medium-scale phenomena. Better results are obtained when predicting widespread rain or snow from cyclonic storm systems. It is anticipated that as the numerical models become more complete in taking into account the relevant atmospheric processes and as observations of the initial conditions become more detailed, the predictions of rain and snow will become more accurate.

By means of a facsimile network, the National Weather Service transmits weather maps depicting the existing state of the atmosphere and predictions of future states. The present-day numerical models do a satisfactory job in predicting the patterns of pressure and wind velocity especially in the middle layers of the atmosphere, such as the 500-mb level. The models still cannot adequately predict surface temperatures, winds, and precipitation. These quantities are significantly influenced by local geographic features as well as the configuration of atmospheric pressure and winds. The presence of a large body of water or hills and mountains affect local weather in ways discussed earlier in this book.

In actual practice, the flow patterns and cloud and precipitation configurations produced by the numerical calculations are used as the first step in weather forecasting. Meteorologists familiar with local influences use the maps produced by the numerical models as a guide for producing specific predictions of temperature, precipitation, and wind.

To an increasing extent, the information on the prognostic charts is used as input data in statistical methods in which the weather in a specific locality is related to the values of pressure, temperature, and humidity at one or more places. Such statistical techniques lend themselves to prediction of the probable occurrence of certain weather events.

For many years the National Weather Service used phrases such as "scattered showers" or "widespread rain" when issuing its daily forecasts.

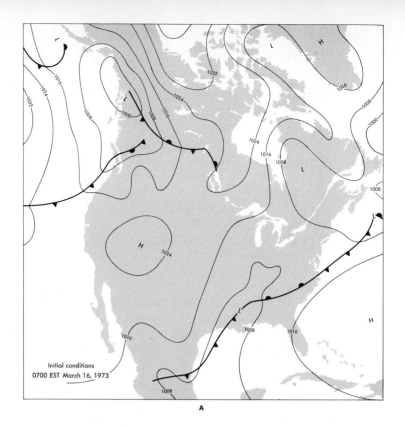

Initial conditions
0700 EST March 16, 1973

A

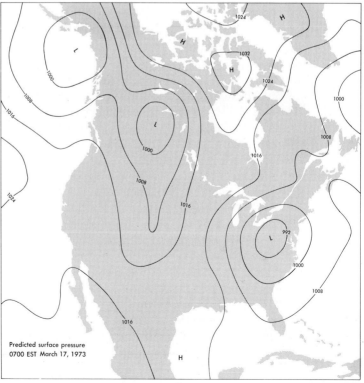

Predicted surface pressure
0700 EST March 17, 1973

B

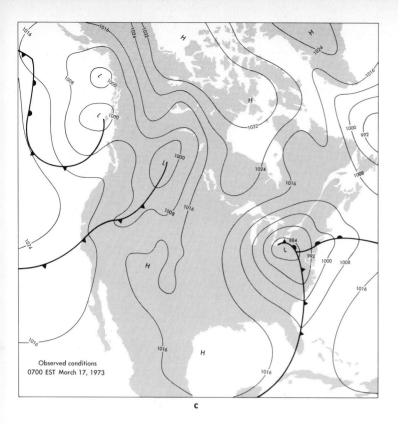

Observed conditions
0700 EST March 17, 1973

C

Figure 12-5 An operational numerical forecast of surface pressure in millibars by means of a mathematical model. (A, *left*) Initial observed conditions at 0700 EST, March 16, 1973; (B, *left*) predicted pattern of surface pressure at 0700 EST, March 17, 1973; (C, *above*) observed conditions at same time as prognostic chart in (B). *Courtesy* F. G. Shuman and J. B. Hovermale, National Weather Service.

But starting about 1965 a more quantitative procedure was adopted, and precipitation forecasts have been expressed in terms of probability. For example, a forecast might read, "The probability of rain today is 30 percent." It means that the chances are three in ten that any random point in the forecast area will experience 0.25 mm or more rain during the daytime. Put another way, this probability forecast means that if there are 10 days when the probability of rain is 30 percent, at your house it should rain on three of those 10 days.

Probability forecasts take into account the nature of the precipitation as well as the confidence of the forecaster. In summer, when showers are the usual form of precipitation, low probabilities are common even when the forecaster is fairly sure some showers will occur somewhere in the vicinity. In the winter, widespread rain or snow is the rule, and higher precipitation probabilities are more common.

Since a rainfall probability forecast is a numerical measure of the likelihood that any locality will be rained on, you can guide your activities in a

12 HOUR ACCUMULATED PRECIPITATION IN MILLIMETERS

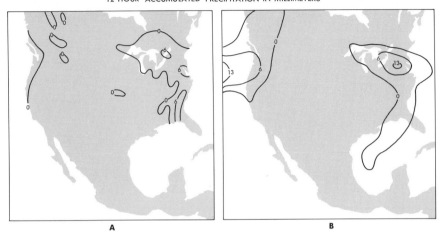

Figure 12-6 Accumulated precipitation 12 hours to 24 hours after initial time: (A) observed; (B) predicted. *From* F. G. Shuman and J. B. Hovermale, *Journal of Applied Meteorology*, 1968, **7**: 525–547.

more rational way than if a less quantitative procedure were used. For example, if you were planning a family picnic, a rainfall probability of 20 percent or less would not be sufficiently threatening to cancel it. The chance of rain at the picnic site is only one out of five and even if it did rain you could pack up and go home. On the other hand, if someone were planning an activity for which rain would be ruinous, such as pouring a great deal of concrete, you would want a day when rainfall probability was very low—perhaps less than five percent. In many occupations it is possible to calculate the losses and gains attributable to various weather phenomena. For example, an electrical utility knows the losses that are incurred when lightning causes power outages, and a construction company can estimate the losses resulting when rain prohibits the pouring of concrete. It can be shown that if reliable probability forecasts are used on a regular basis, many weather-related losses can be reduced.

WEATHER MODIFICATION

In many respects people have been modifying the weather for a long time. In a sense, the construction of a house is a move in that direction, especially when a furnace and an air conditioner are installed. The house allows a person to get out of the rain and snow and to control the temperature and the humidity of the air.

As mentioned earlier farmers modify the growing environment in a variety of ways. By means of trees and bushes in the form of shelterbelts, the damaging effects of wind can be reduced. Techniques of frost protection and irrigation are forms of weather modification.

The weather and climate have been modified inadvertently for a long time. In Chapter 11 it was shown that as cities grow they influence atmospheric conditions, especially temperature and rainfall. The gaseous and particulate pollutants from large industrialized, urban areas may affect the atmosphere over large regions. On a global scale, man-made increases of carbon dioxide and other gases in the atmosphere may change the radiation balance of the earth and possibly influence the global climate. These various possibilities have been discussed earlier.

History is also rich in examples of how people have tried purposefully to change clouds and storms. Little progress in developing procedures for these purposes was made until about 1946. That was when Vincent J. Schaefer, at General Electric's Research Laboratory, showed that Dry Ice could be used to produce ice crystals in supercooled clouds. Shortly thereafter his colleague, Bernard Vonnegut, discovered that certain other substances, such as silver iodide and lead iodide, are effective ice nuclei (Chapter 8).

When a layer of supercooled stratus clouds or fog is seeded either with small particles of Dry Ice or a smoke consisting of silver iodide, huge numbers of ice crystals are produced. They grow rapidly by the ice-crystal process noted in Chapter 8 and in a matter of 5 min to 10 min fall out of the cloud leaving an opening through which the ground can be seen (Fig. 12-7). This scheme is used to clear supercooled fogs over certain airports in the United States, the Soviet Union, and France.

The records reveal, however, that about 95 percent of the fogs occurring in the United States are "warm," that is, they have temperatures above 0°C. Such fogs are the most common variety in most parts of the world. In these cases, ice-nuclei seeding has no effect. Some warm fogs can be dissipated by seeding them with large salt particles, but this scheme is not considered economically feasible. Fogs can be evaporated by heating them. During World War II pipelines with small holes were installed along the runways at 15 airfields in England where fogs are frequent. Aviation fuel was pumped through the pipes and burned as it flamed out of the small holes. The heat effectively evaporated the fog droplets, improved ceiling and visibility conditions, and allowed safe airplane landings. An improved version of this system was tried at the Los Angeles International Airport after the war, but it was not found to be economically effective.

Another technique that uses heat for the dissipation of warm fogs makes use of the hot blasts from jet engines. Some early work was done by U.S. Air Force scientists and engineers in the late 1940's. Further development and testing are still in progress, especially in the United States and

| 10:37 LST at 4909 m | 11:12 LST at 4345 m |
| 11:20 LST at 4909 m | 11:31 LST at 4939 m |

Figure 12-7 When this supercooled altostratus cloud, about 200 m thick, was seeded with Dry Ice, a hole was produced as ice crystals grew and fell from the cloud. *Courtesy* Air Force Geophysics Laboratory.

France. In 1970 a thermal fog dissipation system, called *Turboclair*, was installed at Orly Airport in Paris. Including subsequent modifications in 1972, it consisted of 12 jet engines in underground chambers located along the side of the runway. The objective has been to clear fogs over that part of the runway where landing airplanes touch down. The high-speed jets of hot air produce turbulent air motions. According to French tests, they do not represent a hazard to large commercial or military aircraft, but the turbulence might endanger small aircraft.

Most efforts to change the weather have been directed toward learning how to increase rain or snow. The growing demands for fresh water and the occurrence of periodic droughts, especially in the farming areas of the world, have made the search for new sources of water a vital one.

There still is a great deal of debate among meteorologists and statisticians about the possibility of "rainmaking" by means of cloud seeding. The chief reason for the disagreements lies with the difficulty in evaluating the results of cloud-seeding experiments. For example, after a cloud or a storm system is seeded and the precipitation is measured, there is no way to determine exactly how much rain would have fallen if there had been no seeding. Weather forecasting techniques still cannot make predictions with sufficient accuracy to answer this question. The effects of seeding are expected to be relatively small in relation to the highly variable nature of precipitation. For this reason, it is necessary to use sensitive, well-designed statistical tests that incorporate all that is known about the relevant physical factors. The most satisfactory approach incorporates a randomization procedure whereby only a fraction (usually about one-half) of the suitable clouds or storms are seeded. Then the seeded sample is compared with the nonseeded one, and calculations are made of the likelihood that observed differences were caused by chance rather than by the seeding. If the probability of chance occurrence is found to be small (e.g., less than 5 out of 100), it can be concluded that the observed results were caused by the seeding.

The scientific consensus in the late 1970's was that in certain meteorological circumstances that indicate the presence of clouds of certain types, sizes, and temperatures, ice-nuclei seeding could increase precipitation by perhaps 10 to 30 percent over an area some tens of kilometers in diameter. In other circumstances seeding might decrease precipitation by the same amounts. Under still other conditions, seeding would be expected to have no effect at all. Unfortunately, little progress has been made in identifying the specific meteorological circumstances in terms of expected effects.

Still unresolved is the important question of how far from the "target area" seeding effects are likely to occur. There is some evidence that they may extend more than 200 km downwind of the seeded area, but the arguments for either large-scale increases or decreases of precipitation are not conclusive and need further investigation.

Because of the destructive effects of hail, particularly to vegetation, there have been programs in many countries to develop procedures for reducing the fall of damaging hail. There is a long history of such activity. As long ago as the sixteenth century people were ringing church bells and firing artillery to ward off severe thunderstorms. In 1750 the Archduchess of Austria prohibited the use of guns for hail suppression because of disputes between neighboring landowners over the effects of the firing. Since that time there have been various episodes in which explosives have been used in attempts to reduce hail damage. After World War II farmers in northern Italy began using rockets to carry explosives into thunderstorms threatening to hail on their fruit orchards. It is doubtful that any of these schemes had much effect on the hail.

The more scientifically founded attacks on hailstorms have sought to influence the growth of hailstones. In most of these efforts it has been assumed that the available supply of supercooled water in a hailstorm is essentially fixed. It has been reasoned that if the number of hailstones could be increased, their average size would be reduced. It also has been assumed that the number of hailstones can be increased by introducing a large number of ice nuclei into the supercooled part of the cloud. The aim is to produce hailstones

Figure 12-8 Silver iodide cloud-seeding generators mounted under the wings of a small airplane. A solution of silver iodide in acetone is stored in the tanks (labeled mri). The highly inflammable fluid is fed into the burning cylinder, sprayed by a nozzle, and ignited by a spark. *Courtesy* Meteorology Research Inc.

small enough (less than about 5 mm) to melt as they fall through the warm atmosphere below the level of the 0°C isotherm. In such an event, the precipitation would reach the ground as beneficial rain instead of damaging hail.

Most hailstorm seeding in the United States, Africa, Europe, and Argentina has been carried out from the ground or from airplanes (Fig. 12-8). The results have been mixed and are difficult to interpret. They certainly do not prove that ice-nuclei seeding can diminish the fall of hail. In the Soviet Union silver and lead iodide seeding material has been fired into the supercooled parts of potential hailstorms by means of artillery and rockets. Soviet scientists have reported surprisingly consistent and striking success since the early 1960's. Year after year they have claimed crop damage reduction of 70 percent to 95 percent. Analyses of observations of hailstorms and theoretical studies raise doubts about the validity of the physical model on which the Soviet techniques are based.

As noted in Chapter 9, cloud-to-ground lightning is a serious natural hazard. It kills more people per year than do tornadoes. In addition, lightning causes an average of about 9000 forest and grassland fires every summer in the United States.

Since about the late 1950's there have been several programs aimed at finding ways to reduce the occurrence of lightning and forest fires. Scientists of the U.S. Forest Service conducted experiments in which potential lightning storms were seeded with silver-iodide nuclei. It was hoped that by changing the nature of the cloud and precipitation particles, the rate of cloud charging, and hence the discharge of lightning to the ground, could be reduced. The research produced some encouraging results, but it still has not been convincingly shown that ice-nuclei seeding can reliably reduce the occurrence of the kind of lightning that touches off fires.

Various other schemes for influencing lightning occurrence have been tested in the United States. For example, one of them involves the introduction of millions of short metallic needles into a developing cumulonimbus cloud. They are intended to prevent the buildup of electric charge centers required for the occurrence of lightning.

For reasons noted earlier, hurricanes, especially intense ones that sweep over low-lying coastlines, are very destructive to life and property. Many scientists believe that if the maximum wind speeds in hurricanes could be reduced, there would be a reduction in damage and loss of life. The view that hurricanes might be modified by seeding them with ice nuclei was first advanced by the famous American scientist, Irving Langmuir, who performed a hurricane seeding experiment in 1947.

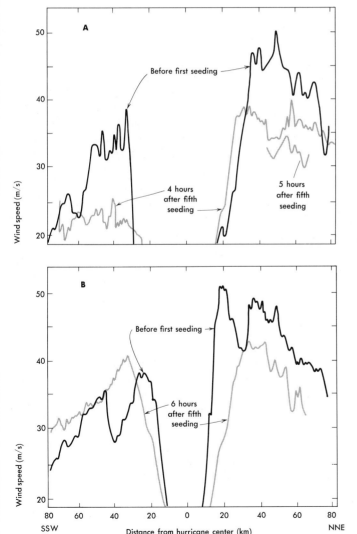

Figure 12-9 Changes of wind speeds with time in hurricane Debbie on (A) August 18, 1969, and (B) August 20, 1969. Winds were measured by an airplane flying at an altitude of 3600 m along a track oriented from south–southwest to north–northeast. *From* R. C. Gentry, *Science*, 1970, **168**: 437 475.

Although there has been much interest in this subject since about 1960, only about half a dozen hurricane seeding tests have been conducted. The most satisfactory ones were carried out on August 18 and 20, 1969, when hurricane Debbie was seeded by aircraft on Project Stormfury, a joint program of the National Oceanic and Atmospheric Administration and the U.S. Navy. As shown in Fig. 12-9, following both periods of seeding, the peak winds within the hurricane decreased substantially at the aircraft flight level of about 3600 m. Mathematical analyses by Project scientists indicated that ice-nuclei seeding of a hurricane outside the zone of maximum winds should cause a reduction of the peak speeds. These experimental and theoretical resutls have led to a feeling of cautious optimism that hurricanes can be weakened. There still are many uncertainties. It is considered essential that more experiments over the open oceans be performed before seeding hurricanes about to strike populated areas.

SOCIETAL CONSEQUENCES OF WEATHER MODIFICATION

It is clear that we still have a great deal to learn about the science and technology of weather modification. It is becoming more evident that as we develop the knowledge and techniques to modify clouds, precipitation, and storm systems, there will be many societal consequences. The United Nations has considered a resolution aimed at preventing the use of weather modification procedures as a weapon. In the United States there have been court actions over such questions as who owns the clouds and precipitation. Many others can be expected to arise. Atmospheric scientists now recognize that the only rational approach to weather modification is one that includes the participation of ecologists, sociologists, legal experts, and the public at large. The goal should be to maximize the benefits for all of society.

APPENDIX ‖

Unit Conversion Factors and Some Properties of the Earth

UNIT CONVERSION FACTORS

Length   1 kilometer (km) $= 1000$ meters (m)

$= 0.6214$ statute mile (mi)

$= 0.5396$ nautical mile (n mi)

1 meter (m) $= 100$ centimeters (cm)

$= 3.281$ feet (ft)

1 centimeter (cm) $= 1000$ millimeters (mm)

$= 0.394$ inches (in.)

1 micrometer (μm) $= 10^{-6}$ m

1 inch (in.) $= 2.54$ cm

1 Angstrom (Å) $= 10^{-10}$ m

Velocity   1 m/sec $= 3.60$ km/hr

$= 2.24$ mi/hr

$= 1.94$ knots

1 knot $= 1$ n mi/hr

$= 1.15$ mi/hr

$= 0.515$ m/sec

Volume   1 m³ $= 10^6$ cm³

1 liter (ℓ) $= 10^{-3}$ m³

$= 1.06$ fluid quarts (qt.)

$= 0.264$ fluid gallons (gal.)

Mass   1 kilogram (kg) $= 1000$ grams (g)

$= 2.205$ pounds (lb)

1 gram (g) $= 0.0353$ ounces (oz)

$= 0.00220$ lb

$$1 \text{ metric ton} = 1000 \text{ kg}$$
$$= 2205 \text{ lb}$$

Pressure 1 standard atmosphere $= 1013.2$ millibars
$$= 14.70 \text{ lb/in.}^2$$
$$= 760 \text{ mm of mercury (Hg)}$$
$$= 29.92 \text{ in. Hg}$$

1 millibar (mb) $= 0.0145$ lb/in.^{2}
$$= 0.750 \text{ mm Hg}$$
$$= 1000 \text{ dynes/cm}^2$$
$$= 100 \text{ newtons/m}^2 \text{ (N/m}^2)$$
$$= 100 \text{ pascals (Pa)}$$

Energy 1 calorie (cal) $= 4.187$ joules
$$= 4.187 \times 10^7 \text{ ergs}$$
$$= 1.16 \times 10^{-6} \text{ kilowatt hour (kWh)}$$
$$= 3.97 \times 10^{-3} \text{ British thermal units (Btu)}$$
$$= 1 \text{ langley-centimeter}^2 \text{ (ly} \cdot \text{cm}^2)$$

Power 1 watt (W) $= 1$ joule/sec
$$= 0.239 \text{ cal/sec}$$
$$= 0.0569 \text{ Btu/min}$$
$$= 0.00134 \text{ electrical horsepower}$$
$$= 2.39 \times 10^{-5} \text{ ly-m}^2/\text{s}$$

SOME PROPERTIES OF THE EARTH

Mass of the earth	$= 5.98 \times 10^{24}$ kg
Mass of the oceans	$= 1.32 \times 10^{21}$ kg
Mass of the earth's atmosphere	$= 5.29 \times 10^{18}$ kg
Mean radius of the earth	$= 6371$ km
Mean distance between the earth and the sun	$= 1.497 \times 10^8$ km
Mean gravitational acceleration at earth's surface	$= 9.807$ m/s
Speed of rotation of a surface point on the earth's equator	$= 460$ m/s
Angular velocity of the earth	$= 7.29 \times 10^{-5}$/s

Humidity Tables

Table A Relative Humidity in Percent as Function of Air Temperature and Wet-Bulb Depression at Atmospheric Pressure Near Sea Level

Dry-bulb temperature (°F)	Wet-bulb depression* (°F)																
	1	2	3	4	5	6	7	8	9	10	15	20	25	30	35	40	45
10	78	60	34	13													
15	82	67	46	29	11												
20	85	70	55	40	26	12											
25	87	74	62	49	37	25	13										
30	89	78	67	56	46	36	26	16	6								
35	91	81	72	63	54	45	36	27	19	10							
40	92	83	75	68	60	52	45	37	29	22							
45	93	86	78	71	64	57	51	44	38	31							
50	93	87	80	74	67	61	55	49	43	38	10						
55	94	88	82	76	70	65	59	54	49	43	19						
60	94	89	83	78	73	68	63	58	53	48	26	5					
65	95	90	85	80	75	70	66	61	56	52	31	12					
70	95	90	86	81	77	72	68	64	59	55	36	19	3				
75	96	91	86	82	78	74	70	66	62	58	40	24	9				
80	96	91	87	83	79	75	72	68	64	61	44	29	15	3			
85	96	92	88	84	80	76	73	70	66	62	46	32	20	8	3		
90	96	92	89	85	81	78	74	71	68	65	49	36	24	13	3		
95	96	93	89	86	82	79	76	72	69	66	52	38	28	18	8		
100	96	93	89	86	83	80	77	73	70	68	54	41	30	21	12	4	
105	97	93	90	87	84	80	78	74	72	69	56	44	34	24	15	8	1

*Wet-bulb depression = dry-bulb temperature − wet-bulb temperature.
Source: C. F. Marvin, *Psychrometric Tables*, Weather Bureau Bulletin 235, 1941.

Table B DEW-POINT TEMPERATURE IN °F AS FUNCTION OF AIR TEMPERATURE AND WET-BULB DEPRESSION AT ATMOSPHERIC PRESSURE NEAR SEA LEVEL

Dry-bulb temperature (°F)	Wet-bulb depression* (°F)															
	1	*2*	*3*	*4*	*5*	*6*	*7*	*8*	*9*	*10*	*15*	*20*	*25*	*30*	*35*	*40*
10	5	−2	−10	−27												
15	11	6	0	−9	−26											
20	16	12	8	2	−7	−21										
25	22	19	15	10	5	−3	−15									
30	27	25	21	18	14	8	2	−7	−25							
35	33	30	28	25	21	17	13	7	0	−11						
40	38	35	33	30	28	25	21	18	13	7						
45	43	41	38	36	34	31	28	25	22	18						
50	48	46	44	42	40	37	34	32	29	26	0					
55	53	51	50	48	45	43	41	38	36	33	15					
60	58	57	55	53	51	49	47	45	43	40	25	−8				
65	63	62	60	59	57	55	53	51	49	47	34	14	−11			
70	69	67	65	64	62	61	59	57	55	53	42	26	15			
75	74	72	71	69	68	66	64	63	61	59	49	36	28	−7		
80	79	77	76	74	73	72	70	68	67	65	56	44	39	19	1	
85	84	82	81	80	78	77	75	74	72	71	62	52	48	32	24	
90	89	87	86	85	83	82	81	79	78	76	69	59	56	43	37	
95	93	93	91	90	89	87	86	85	83	82	74	66	63	52		12
100	99	98	96	95	94	93	91	90	89	87	86	72				

*Wet-bulb depression = dry-bulb temperature − wet-bulb temperature.
Source: C. F. Marvin, *Psychrometric Tables*, Weather Bureau Bulletin 235, 1941.

Climatological Data for Selected Cities

Average monthy temperature and precipitation data for cities in the United States, included in the diagrams in this Appendix, represent averages over the 30-year period ending in 1976. The source of this information is *Comparative Climatic Data Through 1976* (Environmental Data Service, National Oceanic and Atmospheric Administration, Asheville, North Carolina, April, 1977). The data for cities outside the United States are long-term averages obtained from various sources.

It should be borne in mind that even though these are 30-year averages, the specific values plotted on the diagrams depend to a small extent on which 30 years are used. Therefore, if these quantities are compared with so-called normal temperatures and precipitation published elsewhere, differences will be found, but they will be small. One reason for differences in 30-year averages is that the locations of stations are changed from time to time. For example, a weather station positioned at an airport site for 15 years may be changed because of a decision to use the original location for a new building. For this reason, the elevations noted for each city should be regarded as approximate. Information on station locations usually is included in the climatological archives and is available if needed.

The first five pages give climatological data for cities in North America (mostly in the United States) and Hawaii. The last two pages present the data for a few well-known cities on other continents. Information such as this exists for hundreds of cities all over the world. The principal archive for climatological data in the United States is the Environmental Data Service in Asheville, North Carolina.

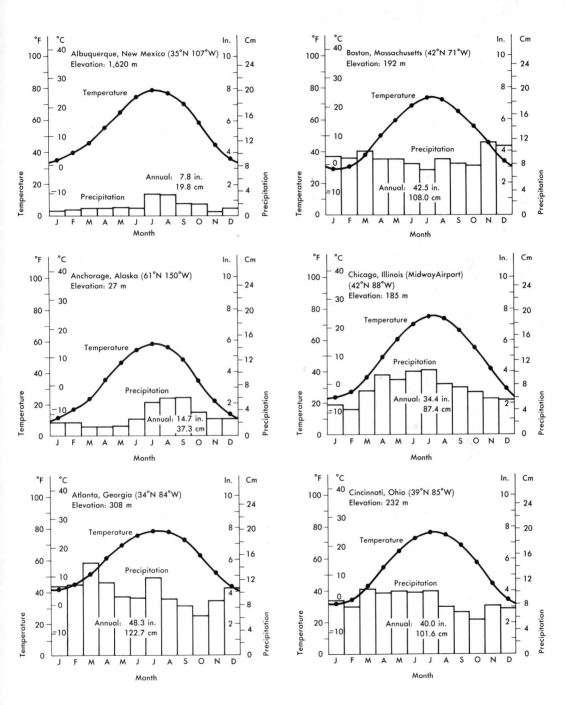

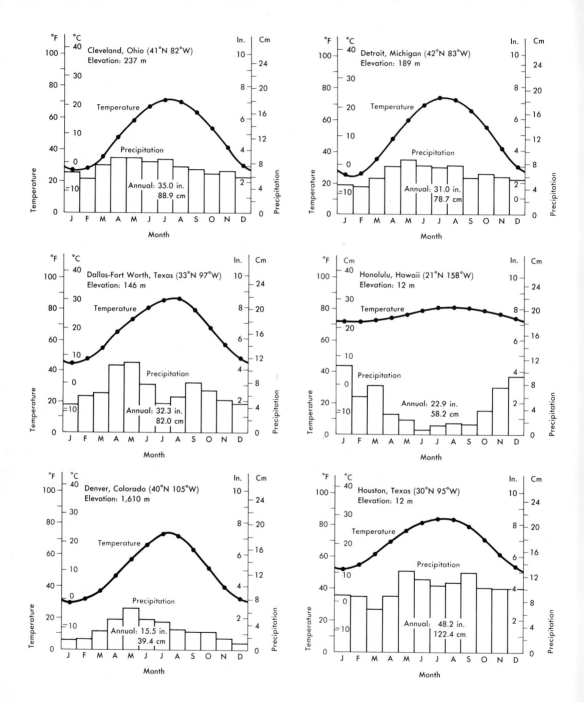

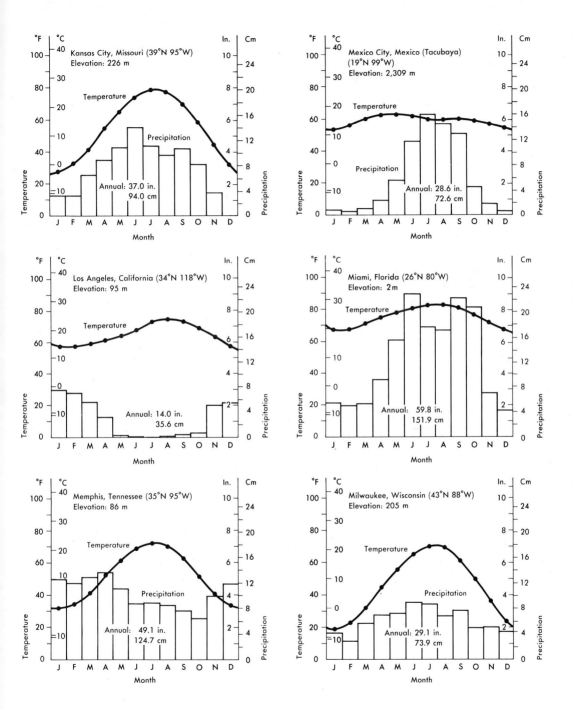

287

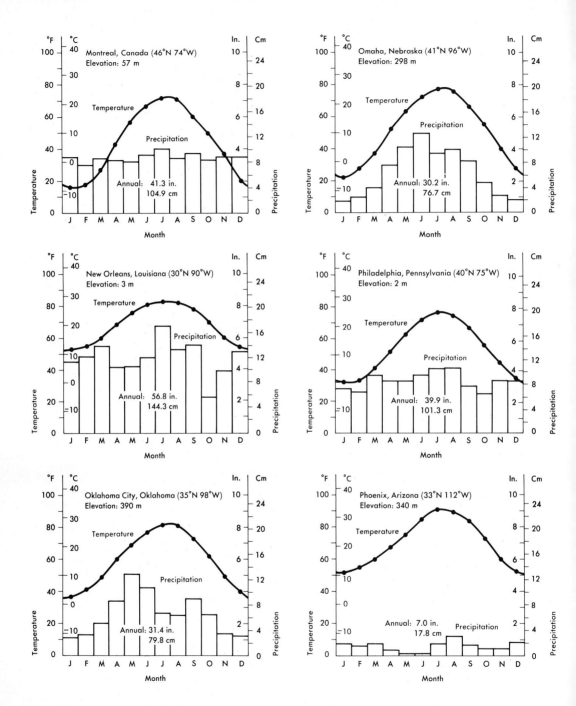

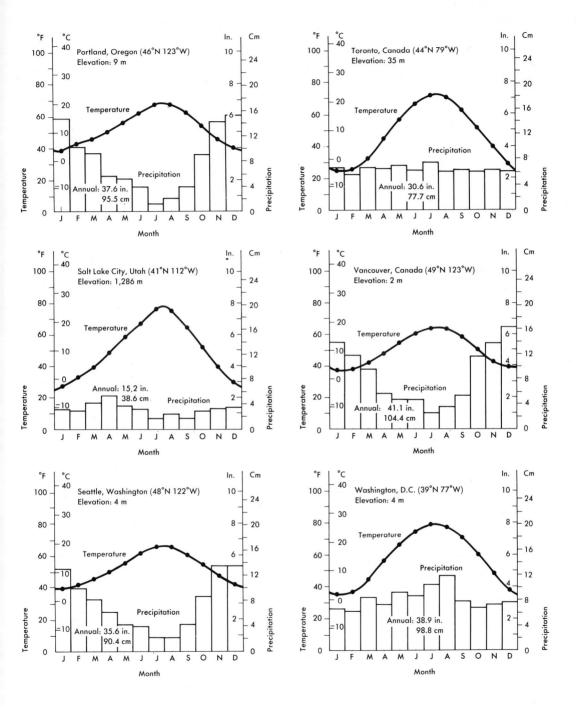

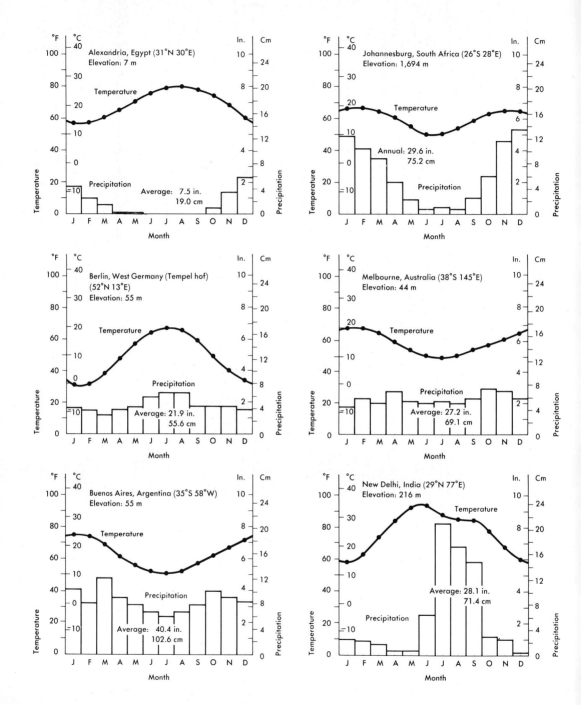

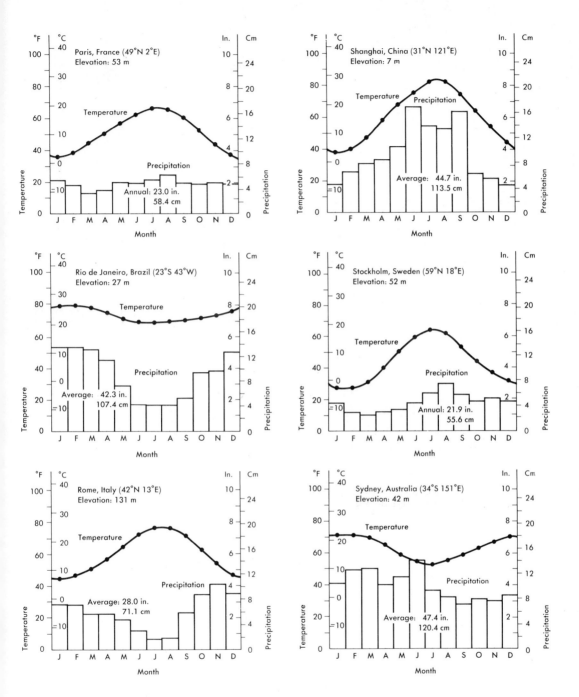

291

Sources
of Further Information

BOOKS
FOR ADDITIONAL INFORMATION

Anthes, R. A., H. A. Panofsky, J. J. Cahir, A. Rango, 1975, *The Atmosphere*, 2nd ed. Columbus, Ohio: Merrill.

Battan, L. J., 1961, *The Nature of Violent Storms*. Garden City: Doubleday.

———, 1966, *The Unclean Sky*. Garden City: Doubleday.

———, 1969, *Harvesting the Clouds: Advances in Weather Modification*. Garden City: Doubleday.

*———, 1973, *Radar Observation of the Atmosphere*. Chicago: University of Chicago Press.

———, 1974, *Weather*. Englewood Cliffs, N.J.: Prentice-Hall.

Bentley, W. A., and W. J. Humphreys, 1962, *Snow Crystals*. New York: Dover Publications.

*Byers, H. R., 1974, *General Meteorology*, 4th ed. New York: McGraw-Hill.

*Craig, R, A., 1965, *The Upper Atmosphere*. New York: Academic Press.

———, 1968, *The Edge of Space*. Garden City: Doubleday.

Dunn, G. E., and B. I. Miller, 1960, *Atlantic Hurricanes*. Baton Rouge: Louisiana State University Press.

*Fleagle, R. G., and J. A. Businger, 1963, *An Introduction to Atmospheric Physics*. New York: Academic.

Flora, S. D., 1956, *Hailstorms of the United States*. Norman: University of Oklahoma Press.

Goody, R, M., and J. C. G. Walker, 1972, *Atmospheres*. Englewood Cliffs, N.J.: Prentice-Hall.

Hess, W. H. (ed.), 1974, *Weather and Climate Modification*. New York: Wiley.

Hidy, G. M., 1967, *The Winds*. Princeton: Van Nostrand.

*For advanced treatment of the subject.

*Holton, J. R., 1972, *An Introduction to Dynamic Meteorology*. New York: Academic.

*Kraus, E. B., 1972, *Atmosphere–Ocean Interaction*. London: Clarendon Press.

Landsberg, H. E., 1969, *Weather and Health*. Garden City: Doubleday.

Lowry, W. D., 1967, *Weather and Life: An Introduction to Biometeorology*. Corvallis: Oregon State University Press.

Ludlam, F. G., and R. Scorer, 1957, *Cloud Study*. London: John Murray.

Mason, B. J., 1962, *Clouds, Rain and Rainmaking*. London: Cambridge University Press.

*———, 1971, *The Physics of Clouds*, 2nd ed. London: Oxford University Press.

Mather, J. R., 1974, *Climatology: Fundamentals and Applications*. New York: McGraw-Hill.

Minnaert, M., 1954, *The Nature of Light and Colour in the Open Air*. New York: Dover Publications.

Ohring, G., 1966, *Weather on the Planets*. Garden City: Doubleday.

Perkins, H. C., 1974, *Air Pollution*. New York: McGraw-Hill.

Riehl, H., 1972, *Introduction to the Atmosphere*, 2nd ed. New York: McGraw-Hill.

*Rogers, R. R., 1979, *A Short Course in Cloud Physics*, Oxford: Pergamon Press.

Scorer, R., 1972, *Clouds of the World*. Harrisburg, Pa.: Stackpole Books.

*Sellers, W. D., 1965, *Physical Climatology*. Chicago: University of Chicago Press.

Trewartha, G. T., 1968, *An Introduction to Climate*, 4th ed. New York: McGraw-Hill.

Tricker, R. A. R., 1970, *An Introduction to Meteorological Optics*. New York: American Elsevier.

*Uman, M. A., 1969, *Lightning*. New York: McGraw-Hill.
———, 1971, *Understanding Lightning*. Carnegie, Pa.: Bek Technical Publications.

*Wallace, J. M., and P. V. Hobbs, 1977, *Atmospheric Science: An Introductory Survey*, New York: Academic.

*White, G. F., and J. E. Haas, 1975, *Assessment of Research on Natural Hazards*. Cambridge, Mass.: M.I.T. Press.

Wilson, C. L., 1972, *Inadvertent Climate Modification: Report of the Study of Man's Impact on Climate (SMIC)*. Cambridge, Mass.: M.I.T. Press.

World Meteorological Society, 1956, *International Cloud Atlas* (Abridged Atlas). Geneva, Switzerland.

FILMS ON METEOROLOGY
(FOR RENT OR PURCHASE)

Above the Horizon (general meteorology)
Formation of Raindrops
Solar Radiation I: Sun and Earth
Solar Radiation II: The Earth's Atmosphere
Atmospheric Electricity
Convective Clouds
Sea Surface Meteorology
Planetary Circulation of the Atmosphere
It's An Ill Wind (about air pollution)

The films listed above are available from Ward's Modern Learning Aids Division, P.O. Box 302, Rochester, N.Y. 14603 or Universal Education and Visual Arts, 100 Universal City Plaza, Universal City, Calif. 91608.

The Inconstant Air (general meteorology)
The Flaming Sky (about auroras)
The Weather Watchers

McGraw-Hill Films, 330 West 42nd Street, New York, N.Y. 10036.

Ice in the Atmosphere

National Center for Atmospheric Research, Boulder, Colo. 80303.

Hurricane
Tornado: Approaching the Unapproachable

Office of Public Affairs, National Oceanic and Atmospheric Administration, Rockville, Md. 20852.

Storms—The Restless Atmosphere
The Atmosphere in Motion
The Ways of Water
Weather Forecasting
Weather Satellites

Encyclopaedia Britannica Films, 425 Michigan Avenue, Chicago, Ill. 60611.

PRINCIPAL JOURNALS AND MAGAZINES
IN THE ENGLISH LANGUAGE

Weatherwise, Heldref Publications, 4000 Albermarle Street, N. W., Washington, D.C., 20016

Science News, 1719 N Street N.W., Washington, D.C. 20036. (Up-to-date news on all sciences including atmospheric sciences.)

NOAA, Office of Public Affairs, National Oceanic and Atmospheric Administration, Rockville, Md. 20852.

Bulletin of American Meteorological Society

*Journal of Atmospheric Sciences

*Journal of Applied Meteorology

*Monthly Weather Review
> These four journals are published by American Meteorological Society, 45 Beacon Street, Boston, Mass. 02108. The asterisks indicate which publications are highly technical.

Weather

*Quarterly Journal of Royal Meteorological Society
> James Glaisher House, Grenville Place, Bracknell, Berks. Rg12 1BX, England.

*Tellus, Wenner-Gren Center, Box 23136, S-10435, Stockholm, Sweden.

WMO Bulletin, World Meteorological Organization, C.P. No. 5, CH-1211, Geneva 20, Switzerland. (Can be ordered from Unipub, Box 433, Murray Hill Station, New York, N.Y. 10016.)

*Journal of Geophysical Research: Oceans and Atmosphere, American Geophysical Union, 1909 K Street, N.W., Washington, D.C. 20006.

*Dynamics of Atmospheres and Oceans, Elsevier, P.O. Box 211, Amsterdam, The Netherlands.

*Climate Change, Reidel Publishing, 306 Dartmouth Street, Boston, Mass. 02116.

Average Monthly Weather Outlook, *Daily Weather Maps*, Climatological data for the states, climatological bulletins, and technical reports. Prepared by National Oceanic and Atmospheric Administration. Write for Price List 48 on Weather, Astronomy, Meteorology. Published by U.S. Government Printing Office, Washington, D.C. 20402.

Weekly Weather and Crop Bulletin, Agricultural Climatology Service Office, South Building Mail Unit, U.S. Department of Agriculture, Washington, D.C. 20250.

SOURCES OF INFORMATION ON ATMOSPHERIC SCIENCES

American Meteorological Society, 45 Beacon Street, Boston, Mass. 02108. (Includes a list of colleges and universities offering programs in meteorology and related sciences.)

American Geophysical Union, 1909 K Street, N.W., Washington, D.C. 20006.

Office of Public Affairs, National Oceanic and Atmospheric Administration, Rockville, Md. 20852.

Public Information Office, National Center for Atmospheric Research, Boulder, Colo. 80303.

Air Pollution Control Association, 4400 Fifth Avenue, Pittsburgh, Pa. 15213.

Unipub, Box 433, Murray Hill Station, New York, N.Y. 10016. (Distributor of reports by World Meteorological Organization.)

METEOROLOGICAL INSTRUMENTS

Science Associates, 230 Nassau Street, Princeton, N.J. 08540.

Weather Measure Corp., P.O. Box 41257, Sacramento, Calif. 95841.

Belfort Instrument Co., 1600 S. Clinton Street, Baltimore, Md. 21224.

Edmund Scientific Co., Edscorp Building, Barrington, N.J. 08007.

Glossary*

Absolute humidity The ratio of the mass of water vapor in a volume of moist air to the volume occupied by the air; also called vapor density.

Absolute temperature The temperature with respect to absolute zero; it is equal to the temperature in degrees Celsius plus 273.

Absorption The process by which radiant energy incident on any substance is retained by that substance and converted into heat or another form of energy.

Acceleration The rate of change of velocity with time. When an object is moving in a straight line, acceleration is change of speed. An object moving along a curved path at constant speed, undergoes an acceleration because there is a time rate of change in the direction of the velocity.

Adiabatic lapse rate The rate of decrease of temperature with height as a volume of air ascends or descends in an adiabatic process. In the earth's atmosphere the dry adiabatic lapse rate is equal to about 10°C/km.

Adiabatic process A thermodynamic change of state of a gaseous system in which there is no transfer of heat into or out of the system. In meteorology an ascending volume of dry air moves to regions of lower pressure, expands, and cools at the rate of about 10°C for every kilometer of ascent. Sinking air is compressed and is warmed at the same rate.

Advection The process of transport of an atmospheric property (such as heat, water vapor, or momentum) solely by the horizontal motion of the atmosphere. In oceanography, advection refers to transport by vertical as well as horizontal motions. *See also* **Convection**.

Aeronomy The science of the upper part of the earth's atmosphere.

Aerosol A physical system made up of solid or liquid particles dispersed in a gas, usually air.

Air mass A widespread body of air that is approximately homogeneous in its horizontal extent, particularly with reference to temperature and moisture.

*Based in part on the *Glossary of Meteorology*, published by the American Meteorological Society, Boston, Massachusetts.

Aitken nuclei Microscopic particles in the atmosphere that serve as condensation nuclei for droplet growth during the rapid adiabatic expansion produced within the instrument called an Aitken dust counter.

Albedo The ratio of the amount of shortwave electromagnetic radiation reflected by a body to the total amount incident on it. The albedo of the earth refers to the ratio of the reflected solar radiation to incident solar radiation.

Altocumulus A middle-altitude cloud type, white or gray in color, made up mostly of water droplets, and occurring in the form of a layer or patches with a waved aspect, the elements of which often appear as rounded masses or rolls.

Altostratus A middle-altitude cloud type in the form of a gray or bluish sheet or layer of striated, fibrous, or uniform appearance. An Altostratus may be made up of water droplets, ice crystals, or both.

Anemometer An instrument for measuring wind speed.

Aneroid barometer An instrument for measuring atmospheric pressure.

Angstrom A unit of length equal to 10^{-10} m and 3.94×10^{-9} in.

Anticyclone A high-pressure system around which the wind blows clockwise in the Northern Hemisphere and counterclockwise in the Southern Hemisphere.

Atmospheric pressure The pressure exerted by the atmosphere as a consequence of the weight of the air lying directly above the unit of area in question.

Aurora The sporadic radiant emissions from the upper atmosphere over middle and high latitudes. They are called *aurora australis* and *aurora borealis* (the northern lights) in the Southern Hemisphere and Northern Hemisphere, respectively.

Ball lightning A rare form of lightning consisting of a reddish, luminous ball about 30 cm in diameter.

Baroclinicity The state of stratification in the atmosphere in which surfaces of constant pressure intersect surfaces of constant density (which depends mostly on temperature). *See also* **Barotropy**.

Barometer An instrument for measuring atmospheric pressure,

Barotropy The state of stratification in the atmosphere in which surfaces of constant pressure and constant density are parallel.

Blackbody A hypothetical "body" that absorbs all of the electromagnetic radiation striking it and radiates the maximum amount of energy at a given temperature.

Blizzard A severe weather condition characterized by low temperatures, strong winds, and a great amount of snow.

Buoyancy The upward force exerted on a volume of fluid (or an object in the fluid) by virtue of the density difference between the volume of fluid (or the object) and that of the surrounding fluid.

Buys Ballot's law   A law describing the relationship of wind direction and the pressure pattern. In the Northern Hemisphere when you stand with your back to the wind, the pressure to the left is lower than to the right. The reverse is the case in the Southern Hemisphere.

Calorie   A unit of heat. The amount of heat required to raise the temperature of 1 g of water 1°C.

Carbon dioxide (CO_2)   A chemical compound composed of one atom of carbon and two atoms of oxygen. It is a heavy, colorless gas whose concentration in the atmosphere is rising. As a solid, CO_2 is called Dry Ice and has a temperature of about −78°C.

Celsius temperature scale   A temperature scale on which the freezing point of water is 0° and the boiling point is 100° at sea level.

Centrifugal force   The apparent force, in a rotating system, deflecting the rotating mass radially outward from the axis of rotation.

Centripetal acceleration   The acceleration of a mass moving in a curved path directed toward the center of curvature. This acceleration is equal to the centrifugal force per unit mass.

Chinook   The name given to the foehn on the eastern slopes of the Rocky Mountains.

Cirrocumulus   A high-altitude, cloud type appearing as a thin, white patch of cloud without shadows and consisting of very small elements in the form of grains and ripples. Although a cirrocumulus is made up mostly of ice crystals, it may contain some supercooled water droplets.

Cirrostratus   A high-altitude cloud type appearing as a whitish veil, usually fibrous but sometimes smooth, which may totally cover the sky. The cloud is made up mostly of ice crystals, and it sometimes produces a halo.

Cirrus   A high-altitude cloud type made up of detached elements in the form of white, delicate filaments, patches, or narrow bands having a fibrous or silky appearance. Cirrus clouds consist of ice particles, sometimes large enough to fall in the form of snow streamers.

Clear air turbulence (CAT)   Turbulence encountered by an aircraft when flying in a region devoid of clouds.

Climate   The statistical attributes of the weather conditions measured over any specified region for a specified interval of time.

Cloud drop or droplet   A spherical mass of water most commonly a few tens of micrometers in diameter formed by the condensation of water vapor. When the drop diameter exceeds 200 μm, it may be regarded as a drizzle drop or raindrop.

Cloud seeding   Any procedure involving the addition to a cloud of certain substances for the purpose of altering the natural development of that cloud.

Coalescence   The merging of two colliding water drops into a single, larger drop.

Cold front   A transition zone between cold and warm air that moves toward the region of warm air.

Condensation   The physical process by which a vapor becomes a liquid or a solid. In meteorology it is used to specify the transformation from gas to liquid.

Condensation nucleus   A particle, either liquid or solid, upon which condensation of water vapor begins in the atmosphere.

Condensation trail (contrail)   A cloud-like streamer often observed behind aircraft flying in clear, cold, humid air.

Constant-height chart   A synoptic chart showing the properties of the atmosphere (e.g. pressure, wind, temperature, etc.) at any specified constant altitude.

Constant-pressure chart   A synoptic chart showing the properties of the atmosphere at any specified constant pressure.

Convection   Motions in a fluid resulting in the transport and mixing of energy and other properties of the fluid. In meteorology, convection usually refers to predominantly vertical motions of a fluid arising because of buoyancy differences. *See also* **Advection**.

Convergence   A state of atmospheric motions that results in a net inflow of air into a specified region; the opposite of divergence. *See also* **Divergence**.

Coriolis force   An apparent force that arises when air velocity is measured with respect to the surface of the rotating earth. The "force" is proportional to the wind speed and acts to the right of the wind velocity in the Northern Hemisphere. Its magnitude increases from zero at the equator to a maximum at the North Pole.

Corona   A set of one or more rings of small radii concentrically surrounding the sun or moon when veiled by a thin cloud consisting of water droplets. When colors are discernible, the primary corona ranges from blue on the inside to red on the outside. *See also* **Halo**.

Cumulonimbus   A cloud type, exceptionally dense and vertically developed, occurring either as isolated clouds or as a line of clouds. A cumulonimbus is made up of water droplets and ice crystals and it usually produces rain (sometimes heavy), lightning, and thunder.

Cumulus   A cloud type in the form of individual, detached elements appearing as rising mounds, domes, or towers with cauliflower-shaped tops; a cumulus is generally dense and has sharp outlines.

Cyclone A low-pressure system around which the wind blows counterclockwise in the Northern Hemisphere and clockwise in the Southern Hemisphere.

Density The ratio of the mass of any substance to the volume it occupies, usually expressed in units of grams per cubic centimeter, grams per cubic meter, or kilograms per cubic meter.

Depression A region of low pressure.

Dew point The temperature to which air must be cooled at constant pressure and water-vapor content in order for saturation to occur.

Diffusion The exchange of fluid substance and its properties between different regions in the fluid, as a result of small, almost random motions of the fluid.

Divergence A state of atmospheric motions that results in a net outflow of air from a specified region. *See also* **Convergence**.

Doldrums The equatorial region characterized by low pressure and light and variable winds.

Drag The frictional resistance offered by air or other fluid to the motion of bodies passing through it.

Drizzle Very small, numerous water drops that may appear to float but are large enough to fall to the ground, usually from stratus clouds, Drizzle drops range in size from 0.2 mm to 0.5 mm in diameter; larger drops are considered to be raindrops.

Drought A period of abnormally dry weather sufficiently prolonged for the lack of water to cause serious deleterious effects on agricultural and other biological activities.

Dry-adiabatic lapse rate *See* **Adiabatic lapse rate**.

Dust devil A small, vigorous whirlwind usually of short duration, rendered visible by dust, sand, and debris. It can rotate either clockwise or counterclockwise.

Eddy A small volume of air within a larger air volume that has certain distinctive properties and a life history of its own.

Electromagnetic radiation Energy propagated in the form of an advancing disturbance in electric and magnetic fields existing in space or in the media through which the wave is passing.

Entrainment The mixing of environmental air into an organized air current, such as an updraft, so that the environmental air becomes part of the current.

Environmental lapse rate The rate of decrease of temperature with height at a given time and place.

Equatorial trough　The zone running roughly east–west that separates the northeast trade winds from the southeast trade winds. Also called the intertropical convergence zone.

Equinox　Either of the two points of intersection of the sun's path and the plane of the earth's equator. The vernal equinox occurs on or about March 21 and the autumnal equinox on about September 22. On these days all over the earth the days and nights are of equal duration.

Evaporation　The physical process by which a liquid is transformed to a gas. In meteorology the transformation from solid directly to gas is called *sublimation.*

Evapotranspiration　The combined processes by which water is transferred from the earth's surface to the atmosphere; evaporation of liquid or solid water plus transpiration from plants.

Exosphere　The outermost portion of the atmosphere whose lower boundary is at a height of about 500 km.

Eye　Usually refers to the roughly circular area of comparatively light winds and fair weather found at the center of a hurricane, typhoon, or similar severe tropical storm.

Fahrenheit temperature scale　A temperature scale on which the freezing point of water is 32° and the boiling point is 212° at sea level.

Foehn　A warm, dry wind on the lee side of a mountain range; the warmth and dryness of the air are the result of an adiabatic compression upon descending the mountain slopes.

Fog　A hydrometeor consisting of a visible aggregate of minute water droplets suspended in the atmosphere near the earth's surface.

Freezing level　The lowest altitude in the atmosphere at which the temperature is 0°C. Note that ascending water drops passing through this level often become supercooled instead of freezing.

Front　The transition zone between two air masses of different air temperature and density.

Frost　A deposit of ice crystals formed by direct sublimation on exposed objects such as grass and tree branches when the air temperature falls below the frost point, that is, when the dew-point temperature is below freezing.

Fusion　The phase transition of a substance passing from the solid to the liquid state, that is, the process of melting.

Geostrophic wind　The horizontal wind velocity resulting when the Coriolis force exactly balances the horizontal pressure gradient force.

Gradient　The rate of decrease of a function in space. For example, the decrease of pressure between two points along a line perpendicular to the isobars divided by the distance between the points is the pressure gradient.

Greenhouse effect The heating of the atmosphere by virtue of the fact that short wavelength, solar radiation is transmitted rather freely through the atmosphere and infrared radiation from the earth is more readily absorbed.

Hail Precipitation in the form of balls or lumps of ice having diameters of 5 mm or more and produced in cumulonimbus clouds.

Halo A class of optical phenomena that appears as colored or whitish rings and arcs around the sun or the moon when seen through a cloud of ice crystals. The most common is the 22° halo, which has an angular radius of 22° and exhibits colors ranging from red on the inside to blue on the outside. *See also* **Corona**.

Haze Fine, dry or wet dust, salt, or other particles dispersed through the atmosphere. The particles are too small to be seen by the naked eye, but they reduce horizontal visibility and give the atmosphere a whitish appearance.

Hertz (Hz) A measure of frequency. One hertz equals one cycle per second.

Heterosphere The layer in the atmosphere above about 80 km throughout which the gaseous composition of the atmosphere changes with height.

Homosphere The layer in the atmosphere below about 80 km throughout which the gaseous composition of the atmosphere is generally uniform.

Hurricane A severe tropical cyclone having a maximum wind speed exceeding 32.6 m/sec (73 mi/hr^{-1}) and occurring over the North Atlantic Ocean, Caribbean Sea, Gulf of Mexico, eastern North Pacific Ocean, and off the coast of Mexico. Similar storms in other parts of the world have different names (e.g., typhoons, cyclones).

Hydrometeor Any water or ice particles formed by condensation of water vapor in the free atmosphere or at the earth's surface or blown by the wind into the atmosphere.

Hydrostatic equation An equation expressing a balance in the vertical between the upward directed pressure gradient force and the downward directed force of gravity. If the vertical distribution of air density is known, the equation allows calculations of pressure as a function of height.

Hygroscopic A property of a substance indicating a marked ability to accelerate condensation. On hygroscopic particles, condensation begins at relative humidities less than 100 percent.

Ice crystal Any one of a number of crystalline forms in which ice appears, for example, hexagonal columns, hexagonal platelets, dendritic crystals, ice needles, and combinations of these forms.

Ice nucleus   Any particle that serves as a nucleus in the formation of ice crystals in the atmosphere.

Infrared radiation   Electromagnetic radiation lying outside the red band with wavelengths between about 0.8 μm and 1000 μm.

Insolation   Solar radiation received at the earth's surface.

Instability   A property of any system such that certain disturbances or perturbations will increase in magnitude.

Internal energy   A form of energy measured by the molecular activity of the system. For a perfect gas, the internal energy is proportional to the absolute temperature.

Intertropical convergence zone (ITCZ)   Same as *equatorial trough.*

Inversion   A departure from the usual decrease or increase with altitude of the value of an atmospheric property. The term usually refers to a temperature inversion through which temperature increases with height.

Ionosphere   The atmospheric shell characterized by high ion density extending from about 70 km to very high regions of the atmosphere.

Isobar   A line of equal pressure.

Isohyet   A line of equal precipitation.

Isotach   A line of equal wind speed.

Isotherm   A line of equal temperature.

Jet stream   Relatively strong winds concentrated within a narrow stream in the atmosphere.

Kelvin temperature   *See* **Absolute temperature.**

Kinetic energy   The energy a mass possesses as a consequence of its motion; it is given by one-half the product of the mass and the square of the speed.

Knot   A unit of velocity equal to one nautical mile per hour.

Langley   A unit of radiant energy falling on a unit area equal to 1 cal/cm².

Lapse rate   The rate of decrease of an atmospheric variable with height, usually referring to temperature.

Latent heat   The heat released or absorbed, per unit mass, by a system during a change of phase. In meteorology at 0°C the latent heats of vaporization, fusion, and sublimation of water are about 600 cal/g, 80 cal/g, and 680 cal/g, respectively.

Lightning flash   The total observed luminous phenomenon accompanying a lightning discharge. It may consist of one or more lightning strokes.

Local winds   Winds which, over a small area, differ from those which would be appropriate for the observed general pressure pattern, usually because of local thermal or orographic effects.

Maritime air A type of air whose characteristics are developed over an extensive water surface and therefore has a high water content in at least its lower levels.

Melting point The temperature at which a solid substance melts and changes to a liquid.

Mesopause The top of the mesosphere corresponding to a level of minimum temperature at about 80 km.

Mesosphere The atmospheric shell between the top of the stratosphere, averaging about 50 km, and the mesopause.

Meteor The streak of light produced by a meteoroid in its passage through the atmosphere.

Meteoroid A small, solid particle from interplanetary space which encounters the earth's atmosphere.

Mirage A refraction phenomenon wherein an image of an object is made to appear displaced from its true position.

Mixing ratio The ratio of the mass of water vapor in a volume of moist air to the mass of dry air in the volume.

Moisture A term in meteorology usually referring to the water vapor content of the air. In climatology, moisture refers to quantities of precipitation.

Momentum The property of a system given by the product of its mass and velocity.

Monsoon A seasonal wind accompanying temperature changes over land and water from one season of the year to another.

Mountain and valley winds A system of diurnal winds along a valley blowing upvalley by day and downvalley by night.

Nacreous clouds Rare clouds occurring at heights of 20 km to 30 km at high latitudes. During daylight hours nacreous clouds have the appearance of pale cirrus; at sunset they become brilliant and colorful. They are also called mother-of-pearl clouds.

Nimbostratus A principal cloud type, gray colored and often dark, and rendered diffuse by more or less continuous rain or show.

Nimbus A term usually designating the presence of rain or snow.

Noctilucent clouds Rare clouds observed at twilight at very high altitudes (between 75 km and 90 km) at high latitudes. They resemble thin cirrus, but they usually have a bluish or silverish color, although sometimes, against a dark sky, they appear orange to red.

Normal In meteorology the term normal is often used to represent the average value of an element such as temperature, presssure, or rainfall over any fixed period of years, for example, 30 years.

Nucleus   In meteorology, a particle of any nature upon which molecules of water or ice accumulate as a result of a phase change to a more condensed state.

Occluded front   A composite of two fronts that is formed as a cold front overtakes either a warm front or a stationary front.

Orographic cloud   A cloud whose form and extent are determined in part by effects of mountains on the air flowing over them.

Ozone (O_3)   An almost colorless (but faintly blue) gaseous form of oxygen, made up of three oxygen atoms and having an odor like weak chlorine.

Ozonosphere   The atmospheric shell in which there is appreciable ozone concentration and in which ozone plays an important part in the radiation balance of the atmosphere. The region lies between about 10 km and 50 km, but maximum ozone concentrations are found at altitudes of 15 to 30 km.

Parcel of air   A small volume of air having uniform properties.

Particulate   Any solid or liquid particle in the atmosphere.

Phenology   The study of how climate affects periodic events in the lives of plants and animals.

Photoionization   The removal of an electron from an atom or molecule by the absorption of radiation.

Planck's law   An expression for the emissive power as a function of wavelength of a blackbody at a given temperature.

Polar air   A type of cold air developed over high latitudes.

Polar front   The semipermanent, semicontinuous front circling the globe and separating polar air masses from warmer tropical air masses.

Potential energy   The energy that a mass possesses as a consequence of its position in the field of gravity. A mass gains potential energy as it is lifted to higher altitudes.

Precipitation   Any form of water particle, whether liquid or solid, that falls from the atmosphere and reaches the ground.

Pressure gradient force   The force arising because of differences in pressure within a fluid. It is proportional to the pressure gradient. Sometimes called the pressure force.

Pressure gradient   The rate of decrease of pressure in space at a given time. *See also* **Gradient**.

Psychrometer   An instrument used for measuring water vapor content of the atmosphere. It is a type of hygrometer consisting of two thermometers, one that is dry and one that has wet muslin over the bulb.

Radiation   The process by which energy is propagated through space in the form of electromagnetic waves.

Radiosonde   A balloon-borne instrument that measures pressure, temperature, and relative humidity and transmits the data to a receiving station.

Rainbow   Any one of a family of circular arcs consisting of concentric colored bands arranged from blue on the inside to red on the outside produced by the refraction and reflection of sunlight by a sheet of raindrops. Sometimes outside the primary bow there is a lighter secondary bow that has red on the inside and blue on the outside.

Raindrop   A drop of water of diameter exceeding 0.2 mm falling through the atmosphere. In careful usage, drops having diameters from 0.2 mm to 0.5 mm are called drizzle drops.

Rawinsonde   A radiosonde tracked by radar or radio techniques in order to obtain wind data in addition to pressure, temperature, and relative humidity.

Reflectivity   The ratio of radiation energy reflected to the total radiant energy incident upon a surface.

Refraction   The process in which the direction of radiant enerty propagation is changed as a result of changes in the physical properties within the propagating medium or as the radiation passes from one medium to another.

Relative humidity   The ratio, expressed in percent, of the amount of water vapor in the air to the amount at saturation vapor. The amount of water vapor preferably should be measured in terms of vapor pressure, but it can be measured in terms of absolute humidity, mixing ratio, or specific humidity.

Saturated air   Moist air in a state of equilibrium with a plane surface of pure water or ice at the same temperature and pressure. In such a state the relative humidity is 100 percent and the amount of water vapor is maximum for the given temperature.

Saturation-adiabatic lapse rate   The rate of decrease of temperature with height as a volume of saturated air is lifted adiabatically. Because of the release of latent heat, this lapse rate is less than the adiabatic lapse rate.

Scattering   The process by which small particles suspended in a medium diffuse, in all directions, a portion of the incident radiation.

Sea breeze   A coastal local wind that blows from sea to land; it is caused by temperature differences when the land is warmer than the adjacent sea surface.

Sferics   A contraction of the word "atmospherics," which refers to the radio signal produced during an electric discharge such as lightning.

Shelterbelt   A belt of trees, shrubs, or artificial barriers, arranged as a protection against strong winds.

Sleet   Small pellets of transparent or translucent ice 5 mm or less in diameter. In British terminology, precipitation in the form of a mixture of rain and snow.

Smog   A natural fog contaminated by industrial pollutants; a mixture of smoke and fog.

Snow   Precipitation made up of white or translucent ice crystals, chiefly in complex branched hexagonal form and often agglomerated into snowflakes.

Snow pellets   Precipitation consisting of white, opaque approximately spherical ice particles having a snow-like structure and about 2 mm to 5 mm in diameter.

Solar constant   The rate at which solar radiation is received outside the earth's atmosphere on a surface perpendicular to the incident radiation at the earth's average distance from the sun.

Solstice   Either of the two points in the sun's path when it is displaced farthest north or south of the earth's equator. The summer and winter solstices occur on about June 22 and December 21, respectively.

Specific heat   The quantity of heat that must be absorbed (or released) by unit mass of a system to cause a 1°C temperature rise (or fall).

Specific humidity   The ratio of the mass of water vapor in a volume of moist air to the total mass of the volume of moist air.

Standard atmosphere   A hypothetical vertical distribution of atmospheric temperature, pressure, density, and other properties which, by international agreement, is taken to be representative of the atmosphere for one or more specific purposes.

Stationary front   A front that does not move or moves very little.

Stefan-Boltzmann law   A law that states that the amount of energy radiated per unit time from a unit area of an ideal blackbody is proportional to the fourth power of the absolute temperature of the blackbody.

Storm surge   An abnormal rise of the sea along a shore as a result, primarily, of the winds of a storm.

Stratocumulus   A principal cloud type in the form of gray or whitish layers or patches; a stratocumulus almost always has dark parts and it is nonfibrous.

Stratosphere   The atmospheric shell above the troposphere and below the mesosphere; it extends from about 10 km to 50 km, on, the average, above the earth's surface.

Stratus   A principal cloud type in the form of a gray layer with a rather uniform base.

Sublimation   In meteorology the transition of a substance from the solid phase directly to the vapor phase, or sometimes *vice versa*, without passing through the liquid phase.

Subsidence   A descending motion of air, usually with the implication that the condition exists over a broad area.

Subtropical high   One of the commonly observed large high-pressure regions distributed around the earth, centered at about latitudes 30°N and 30°S.

Superadiabatic lapse rate   An environmental lapse rate greater than the dry-adiabatic lapse rate.

Supercooled cloud   A cloud made up of liquid droplets at temperatures below 0°C.

Supersaturation   The condition existing in a volume of air when the relative humidity exceeds 100 percent, that is, when the air contains more water vapor than is needed to produce saturation over a plane surface of pure water or ice.

Suspensoid   A system consisting of one substance (such as dust) dispersed through another (such as air) in a moderately finely divided state.

Synoptic chart   Any chart or map on which data and analyses are presented to describe the state of the atmosphere over a large area at a given moment in time.

Temperature inversion   A layer through which air temperature increases with altitude.

Terminal velocity   The constant falling speed, through any fluid medium, at which the upward drag and buoyancy forces just equal the downward gravitational force on an object.

Thermosphere   The atmospheric shell extending from the top of the mesosphere at about 80 km to outer space; it includes the exosphere and the ionosphere.

Thunderstorm   A relatively small storm involving one or more cumulonimbus clouds, always accompanied by lightning and thunder, generally with rain and gusty winds, and sometimes with hail.

Tornado   A violently rotating column of air extending downward from a cumulonimbus cloud usually having the appearance of a funnel, a column, or a rope.

Trade winds   The winds that occupy most of the tropics and blow from the subtropical highs toward the equatorial trough; they are mostly northeasterly in the Northern Hemisphere and southeasterly in the Southern Hemisphere.

Transpiration   The process by which water in plants is transferred as water vapor to the atmosphere.

Tropical cyclone   A closed, low-pressure center originating over tropical oceans. Tropical cyclones are classified according to their peak wind speeds: *tropical depressions*, with winds less than 17.4 m/s (39 mi/hr); *tropical storms*, with winds 17.4 m/sec⁻¹ to 32.6 m/s (39–73 mi/hr); *hurricanes*, with winds exceeding 32.6 m/s.

Tropopause   The boundary between the troposphere and the stratosphere. Its height varies from 15 km to 20 km in the tropics to about 10 km over the poles.

Troposphere   The lowest 10 km to 20 km of the earth's atmosphere characterized by decreasing temperature with height.

Turbidity   Any condition of the atmosphere that reduces its transparency to radiation, especially visible radiation.

Turbulence   A state of fluid flow exhibiting irregular fluctuations of a nearly random nature.

Typhoon   A severe tropical cyclone in the western Pacific; same as a hurricane.

Ultraviolet radiation   Electromagnetic radiation at wavelengths shorter than visible radiation (0.4 μm) but longer than X rays (about 0.01 μm).

Upwelling   The rising of water toward the surface from subsurface layers of a body of water.

Valley wind   A wind blowing upvalley during the day.

Vapor   Any substance existing in the gaseous state.

Vapor pressure   The pressure exerted by the molecules of a given vapor.

Virga   Wisps or streaks of water and ice particles falling out of a cloud but evaporating before reaching the earth's surface.

Visibility   The greatest distance at which it is possible to see and identify with the unaided eye (a) in the daytime a prominent dark object against the sky at the horizon, and (b) at night a moderately intense light source.

Warm front   A front that moves in such a way that warmer air displaces colder air.

Waterspout   A column of violently rotating air over water having a similarity to a dust devil or tornado.

Weather modification   In general, any effort to alter by artificial means the natural phenomena of the atmosphere.

Weather satellite   An orbiting satellite used to measure the properties of the atmosphere.

Westerlies   The dominant west-to-east motion of the atmosphere centered over the middle latitudes of both the Northern Hemisphere and the Southern Hemisphere.

Wet-bulb temperature The temperature a volume of air would have if cooled, at constant pressure, by evaporation as water is evaporated into the air until it is saturated. It is also the temperature read by a wet-bulb thermometer, that is, one having its bulb covered with wet muslin and ventilated until it reaches a minimum temperature.

Wien's law A law stating that the wavelength of maximum radiation intensity for a blackbody is inversely proportional to the absolute temperature of the blackbody.

Wind rose A diagram designed to show the distribution of wind direction experienced at a given location over a considerable period.

Wind vane An instrument used to indicate wind direction.

X ray Electromagnetic radiation of very short wavelength lying within the interval of 0.01 μm to 0.0001 μm.

Index

Absolute humidity, 93–95
Absorption (*see also* Absorptivity)
  of solar radiation, 47–49
  of terrestrial radiation, 46–47, 255–257
Absorptivity, 44, 46–49
Acceleration, 65–66
Acid particles, 19
Acid rain, 19
Adiabatic processes and lapse rates:
  dry, 85–93
  moist, 93–100
Aerosols:
  air mass characteristics, 127–133
  in the atmosphere, 19–23, 91–93, 162, 254–256
  definition, 11
  effects on climate, 120, 254–256
  residence times, 22, 88–89
  from smoke stacks, 91–93
Aerovane, 64–65
Agriculture, 78–79, 106, 183, 225–228, 241, 243,
    260–263, 276
Agui Current, 233
Air Force Cambridge Research Laboratories, 36
Air Force Geophysics Laboratory, 274
Air Masses:
  classification, 127
  definition, 125
  formation and characteristics, 127–133
Air parcel, 83
Air pollution:
  and climate change, 255–256
  diffusion in the atmosphere, 91–93
  gaseous pollutants, 17–19
  particles (*see* Aerosols)
  stratospheric, 111
  trapping in stable air, 127–131
Air Pollution Control Association, 297
Aitken particles, 21–22, 162
Albedo:
  of earth, 51
  of various surfaces, 58–59

     *Also see Glossary (pages 297–311) for definitions

Alberta low, 143
Aleutian low, 105, 107–108, 143
Altocumulus, 149, 151, 152
Altostratus, 151, 155
American Geophysical Union, 296
American Meteorological Society, 296, 298
American Red Cross, 204
Anemometer, 64
Angell, J. K., 253
Annual changes:
  of precipitation, 235–242, 286–291
  of temperature, 230–234, 286–291
Anthes, R. A., 293
Anticyclones:
  pressure and winds, 70–73
  scales, 102
  semipermanent:
    evaporation under, 177–179
    general circulation, 89–90, 103–105, 107, 245, 248
Aphelion, 40
Arakawa, A., 119
Atacama Desert, 105
Atmospheric electricity:
  and coalescence, 168
  and the ionosphere, 35–37
  in a thunderstorm, 186–187
Aurora, 37–38
Aviation and weather:
  affects of fog, 265, 273–274
  flight planning, 111, 261, 263
  turbulence, 111–112, 133, 185, 263–264
Azores high, 105

Baroclinic:
  atmosphere, 138
  numerical models, 268
  theory of cyclones, 141–142
Barograph, 25–27
Barometer, 25–26
Barotropic atmosphere, 138